W9-CPZ-920

THE ILLUSION OF TRUST

Theology and Medicine

VOLUME 5

The titles published in this series are listed at the end of this volume.

THE ILLUSION OF TRUST

Toward a Medical Theological Ethics in the Postmodern Age

by

EDWIN R. DuBOSE

The Park Ridge Center for the Study of Health, Faith, and Ethics
Chicago, Illinois, USA

KLUWER ACADEMIC PUBLISHERS
DORDRECHT / BOSTON / LONDON

Library of Congress Cataloging-in-Publication Data

DuBose, Edwin R.
The illusion of trust : toward a medical theological ethics in the postmodern age / by Edwin R. DuBose.
p. cm. -- (Theology and medicine ; v. 5)
Includes bibliographical references and index.
ISBN 0-7923-3144-3 (hb : alk. paper)
1. Medical ethics--Religious aspects. I. Title. II. Series.
[DNLM: 1. Ethics, Medical. 2. Physician-Patient Relations.
3. Religion and Medicine. W1 TH119M v. 5 1995 / W 50 D817i 1995]
R725.55.D83 1995
174'.2--dc20
DNLM/DLC
for Library of Congress 94-34129

ISBN 0-7923-3144-3

Published by Kluwer Academic Publishers,
P.O. Box 17, 3300 AA Dordrecht, The Netherlands.

Kluwer Academic Publishers incorporates
the publishing programmes of
D. Reidel, Martinus Nijhoff, Dr W. Junk and MTP Press.

Sold and distributed in the U.S.A. and Canada
by Kluwer Academic Publishers,
101 Philip Drive, Norwell, MA 02061, U.S.A.

In all other countries, sold and distributed
by Kluwer Academic Publishers Group,
P.O. Box 322, 3300 AH Dordrecht, The Netherlands.

Printed on acid-free paper

Printed in the Netherlands

TABLE OF CONTENTS

PREFACE

This book is about trust and its implications for a medical theological ethics. Beginning with its earliest work, there has been attention to trust running through the bioethics literature in the United States, and much of this discussion has examined its theological elements. Clearly, trust is indispensable when describing the patient-physician relationship, so why is there a need for yet another study?

There is no doubt that people generally trust physicians. Traditionally the physician is the patient's fiduciary agent, whose sole obligation is to act only in the patient's best interest. In recent times, however, there is a perception on the part of people within and without health care that physicians have other obligations that compete with their obligation to the patient. If we acknowledge that one price for the successes of technological biomedicine is high in terms of financial cost, another price of success seems to be distrust, cynicism, and suspicion directed by the public toward the medical profession. If this uneasiness is the price society pays for medical success, what is the price of success for the doctor?

Because of their role within the social order, physicians have claimed and been granted autonomy, authority, and special status. In return, the profession has pledged to serve the well-being and interests of humankind. This fiduciary commitment becomes a taken-for-granted aspect of the physician's identity, both for the physician for whom this dedication is definitional and for the public which expects trustworthy service from this person. With the social and economic changes affecting health care in recent years, however, suspicions have been roused regarding the depth of the profession's fiduciary commitment. Although this suspicion is manifest in a variety of forms, in its most common (and, therefore) disturbing expression, it disrupts the trusting relationship between physician and patient on which much of the caring and curing within "health care" depends. Efforts to counter suspicion and distrust, it seems to me, tend to exacerbate the problem. As a result, the taken-for-granted assumption on the public's part that physicians are trustworthy is eroding, and there is a corresponding effect on the way in which physicians understand and value their work. As we move into an era of health care reform, the ways in which trust operates within the doctor and patient relationship will shape the economic and social structures that govern the clinical encounter. With all the talk of "security" as a value driving health reform, it is well to review the value of the medical profession's fiduciary commitment.

Research and writing is a very personal experience. At the same time, this undertaking would not have been possible without the willing help and constant support of many people to whom I am indebted. Among these are Niels C. Nielsen and James Sellers; my colleagues at the Park Ridge Center for the Study of Health, Faith, and Ethics; and especially Earl E. Shelp for his friendship and patient guidance. Finally, my greatest appreciation goes to Alice for her support and encouragement, and for knowing how much this work has meant to me.

CHAPTER 1

THE PROBLEM AND THE PURPOSE

It is axiomatic that trust is essential to social relationships, certainly to the clinical encounter that lies at the heart of medical care. Eric Cassell observes that a sick person is not a well person with a disease, one who functions in the everyday world with the addition of just one more element. The taken-for-granted experience of everyday life is transformed for the sick. The power of sickness, Cassell says, is its ability to disconnect a person from the real world [40]. Normally we are connected with the world through our senses, which, along with other connections, tell us we are alive. But when we are sick, however slightly, some of these connections are lost or distorted, and our sense of control, of being at ease in the world, is shaken. But that is what the experience of illness is: our world is changed, it is not what it once was. Distress, pain, and weakness become boundaries of experience, creating the need for support, comfort, and help from others. Yet this need is tinged with an edge: the sick are confronted by their neediness, their dependence on others. It demonstrates their vulnerability.

It is no wonder that trust appears most precarious, and most necessary, at those times when our vulnerability, our sense of dislocation, is the greatest. In health care, the primary function of the care giver toward the one seeking care is to defuse a stark confrontation with finitude, from feelings of helplessness and dislocation that occur when illness casts one out of everyday life and deposits one in a different place ([80], p. 31).[1] Because being ill creates dislocation and dependency, a patient is drawn to trust the knowledge and skills of physicians and their commitment to support rather than exploit his or her vulnerability. There are three elements supporting the medical profession's traditional claim to the public's trust. Physicians, it is said, can be trusted, first, because they possess the formal knowledge and skill that, second, enables them to act in our best interests. Third, their commitment to one and two is inherent in the nature of their profession. Public acceptance of this claim has given physicians great power and authority in our society.

Social power and authority, however, always rest on illusion. Physicians can claim knowledge, competence, and compassion in their work, and they may very well deserve the public's respect for the immense dedication required in the work they do. However, their power, authority, and prestige reside in the image cultivated by physicians that they know what they are doing, and that they know what is best. It is no longer clear that physicians know what they are doing or know what is best for their patients. As a result, many people are growing distrustful of physicians; increasingly, the public does not take for granted the profession's traditional claims to authority and control in its work.

Thus, there is a curious ambivalence on people's part regarding physicians. On the one hand, there is a growing sense of questioning and dissatisfaction, an uneasiness, with the modern medical model. On the other hand, given the obvious successes of scientific medicine, people continue to seek, in fact demand, physicians' care.

The nature of illness and health care makes trust a basic ingredient in the clinical encounter between patients and clinicians, but changes in the cultural and social framework within which medical relationships exist, and of which they are a part, are affecting the way trust shapes the nature of that encounter.[2] Not so long ago, certainly in the days before penicillin, physicians had limited resources to offer. As Lewis Thomas recalls, in the days when the black bag contain little in the way of medical armamentaria, by necessity a doctor functioned more as an advisor or counselor, one who could not do much more than support the patient through the crisis ([269], p. 15). Without question the most important person in the transaction was the patient, whose willing participation and confidence in the plan for recovery largely determined the outcome. Now, Paul Brand writes, in the patient's view at least, the tables have turned: the patient tends to regard the physician as the important party, mainly due to the technical and pharmaceutical advances of recent decades ([22], p. 240).

Medicine has become so complex and elitist that patients may feel helpless, and doubt whether they have much to contribute to their treatment. Too often the patient sees himself or herself as a victim, someone for the experts to pick over, not a partner in recovery and health. In the United States, advertising feeds the victim mentality by conditioning us to believe that health and health care is a complicated matter far beyond the grasp of the average person. We get the impression that, were it not for "vitamin supplements, antiseptics, painkillers, and a trillion-dollar annual investment in medical expertise, our fragile existence would soon come to an end" ([22], p. 240). The development of patients rights, consumerism in health care, the rise of bioethics, and the drive for health care reform may be a reaction to this victim mentality, an effort to redress a balance of power between patient and physician, but it is not coincidental that these trends have accompanied dramatic changes in medical knowledge and expertise. As we enter the mid-1990s, it seems as if increases in medical knowledge, and power, correlate with a lessening of the taken-for-granted belief that physicians can be counted on to act in a fiduciary manner.

In the last twenty-five years, we have moved from a positivist, rational, and problem-solving focus in medicine to a sense that things are out of sorts. Medicine, like other American professions, has suffered a stunning loss of public trust since the 1970s. With relative suddenness medicine has become a focus for debate and ambiguity, if not skepticism and hostility about physicians' economic motives and dedication to their patients' interests ([23], p. 207). Coupled with general concerns about rising health care costs, physicians now often are perceived as functioning "with the business ethic rather than the professional ethic" ([170], p. 2879). Also, rather than engendering trust, the success of technological medicine in the last generation often raises people's anxieties about their ability to make choices for themselves. Increased medical reporting raises public expectations of success in medical matters, yet it also exposes the dilemmas caused by medical progress. Our expectations of medicine are accompanied by a sense of the gap between professional and public knowledge, exacerbating anxieties about our vulnerability in clinical encounters. A demand for autonomy in decision-making contributes to an uneasiness with professional privilege and power. Public concern with controlling medical expenditures and professional power is redrawing the de facto secular "contract" between the medical profession and society, subjecting medical care to the discipline of politics or markets, and reorganizing its basic institutional structure ([262], p. 380).

There are also philosophical challenges to the medical profession's traditional claims to trustworthiness. In fact, as Robert Veatch argues in a recent article, it is conceptually impossible, according to contemporary philosophy of science, to present "value-free facts" to the client. In addition, there are good reasons why professionals ought not to be able to know what their clients' best interests are. Finally, we cannot take for granted the traditional notion that professionals can be trusted to act on a univocal set of virtues inherent in the profession, or that there is a single, definitive conceptualization of how the profession ought to be practiced ([282], pp. 160–161). If these traditional fiduciary claims associated with the medical profession are called into question, how meaningful in the clinical encounter is talk of trust when one no longer assumes that physicians naturally deserve it?

The time seems ripe for a new understanding of the way trust operates in the clinical encounter. If the basic ingredient in clinical relationships can no longer be taken for granted, then fiduciary language as a basis for the physician-patient relationship is called into question. The way in which trust operates (or functions) to create, sustain, and modify social relationships is a complex phenomenon that is difficult to define and distinguish. Trust is confidence in or reliance upon someone or something without investigation or evidence ([45], p. 180). Therefore, it is always accompanied by an element of risk and ambiguity. In spite of the risk, or more accurately, because of it, the person or thing in which trust is reposed becomes a locus of belief, expectation, and hope.

Trust also represents a medium of social exchange, operating on several levels, which serves to regulate our relations with each other ([218]). By reducing the complexity of social situations, especially as this complexity results from other people's freedom, trust shapes our expectations of others and of the situations in which we find ourselves. Ordinarily we assume that a person will act consistently with the image or personality that he or she has made socially visible. However, as Bernard Barber argues, trust is not a function of individual personality variables nor of abstract moral argument, but a phenomenon of social and cultural variables ([8], p. 5). We may trust someone's knowledge or character, but our expectations of that person also are shaped by social institutions and cultural values. Therefore, broad social changes affect the meaning and interpretation of trust, with a corresponding effect on interpersonal relationships.[3] In the absence of trust, people seek more direct control over their relationships, and, at least on an immediate emotional level, contract language offers some measure of control, some way to limit the ambiguity of social interactions. Under these conditions, however, trust becomes a bargaining chip in relationships that develop to protect the parties, one from the other. Distrust becomes the functional equivalent of trust in social relationships ([8], p. 22).

The reinterpretation of self-protection and its effect on the fiducial component of professional identity has implications for a medically informed theological ethics. Social interaction necessarily involves relationships of power. Biblically, the poor, the sick, and the vulnerable are to be protected from exploitation. Their presence within a society represents a call to communality and solidarity, a limitation to self-interest.[4] The relationship between the medical profession and the public, particularly the specific dimension of the physician-patient encounter, is suited to theological investigation because of the issues of the dependency and vulnerability caused by illness and the possibility for human exploitation.

Theological understandings of bioethics emphasize a sense of community and enunciate an ethic of giving and receiving, of caring, of helping, and a medical moral responsibility premised on fiduciary relations between physicians and patients, and between the medical profession and society ([255], p. 2). Ramsey, for example, wrote of reciprocal trust between doctors and patients grounded in a cooperative agreement of covenant-fidelity, as that which allows the healing relationship to occur [229]. According to him, the physician's authority is generated by the patient's free choice and by the mutual trust and faithfulness of the relationship. However, a generation later, with a turn towards individualism and the absolutization of autonomy in health care, a relationship of mutual trust and faithfulness seems the exception, rather than the rule.[5]

My aim is to prepare the soil for a theologically minded ethics of the modern medical profession. During the last twenty years, reflection on ethics in health care has been dominated by philosophers, but recently growing dissatisfaction with the state of bioethics has led to renewed theological work. This work has focused on notions of the common good, dimensions of the covenant, and conceptions of character and virtue. However, the very theological strength of these appeals to transcendent possibilities for human vision and action reduce the value of their prescriptions as vehicles for professional transformation ([129], p. 111). They fail to take adequate account of the elements of finitude – the self-referential viewpoints and interested rationalities – which constrain human thought and action in the professional setting.

Ethicists who articulate visions of the common good, covenantal ideals, and virtues as an expression of faith helpfully challenge physicians to cultivate broad and transcendent loyalties. However, they tend to ignore the prior and foundational question of how a trust expressive of a fundamental trust in God can be established and sustained in the clinical setting. We need to begin with an account of how physicians' expectations of their work are shaped by the limits of their profession. From such an account, we can explore ways in which physicians (and patients) might come to trust the environment of their interaction, as a presupposition for developing more comprehensive loyalties ([129], p. 112).

It is important to appreciate the fact that we must first investigate the social environment of trust before we can address the theological implications of trust for medical ethics. The first prescriptive task of theological ethicists is not to call physicians to cultivate an allegiance to higher values or virtues, but to explain how they might develop a fundamental trust, constituted in part by proximate relations of trust, in the clinical setting. The point is not that theologically-inflected ethics of aspiration, vision, and virtue are necessarily irrelevant or inappropriate, but that they are valid – as a priority in prescription – only 1) when they rest upon an understanding of how physicians are shaped by the modern medical profession's understanding of the limits of their work; and 2) when the prior problem of the possibility of trust has been addressed ([129], p. 112). Therefore, we must examine why attention to an erosion of trust between the medical profession and the public is important in any effort to revitalize medical ethics.

It commonly is assumed that people trust physicians; it is taken for granted that physicians see themselves as trustworthy. However, there are recent signs that serve to call the first assumption into question. I am not interested in the reasons why public confidence in physicians has been eroding. I am interested in the effect that this erosion has on the physician's perception of himself or herself as a physician whose identity is tied to a declaration of fiduciary commitment to his or her patients. In reaction to an erosion of confidence in its image, the profession appeals to the physician's dedication to superior knowledge balanced by a fiduciary commitment to service as virtues characterizing the "good" physician.[6] However, as the physician's medical expertise and knowledge have grown, the understanding of trust as a virtue in relations between patient and physician has undergone a shift. The protestation of trust, the declaration of trustworthiness, by the profession becomes an illusion masking an uneasiness felt by the profession's members in their relations with others.

According to sociological theory, human society is a complex matrix of associations and relations which results from the social negotiations among individuals and institutions. Ordinarily, we assume that others present themselves to us in a trustworthy fashion, and that negotiations are based upon the desire to achieve relations of mutual benefit. When the balance of power among individuals and social groups is upset, the parties seek to restore an acceptable balance of power. During these periods, each party becomes defensive and self-protective in its relations with the other, each striving to preserve its power and autonomy in what has become an adversarial relationship. In such a relationship, trust can be used as a manipulative tool to enhance the power and control of one party in the social matrix.[7] The fiduciary element in the relationship becomes a shell, outwardly polished but inwardly hollow. This condition is the result of a serious misunderstanding of the distinctions between confidence, trust, and faith.

Often these terms are used interchangeably. I can have confidence in the physician's skills and knowledge, but I may not trust him with my welfare. The application of the phenomenological method to the profession's fiduciary focus uncovers two dimensions to trust. The first dimension allows me to take for granted that the physician possesses the requisite knowledge and skills, and to posit confidence in the assertion that he or she is dedicated to my interests. This first instance of trust is equivalent to the phenomenological epoché of the "natural attitude" in which a person assumes reality is as it is represented. In a climate of eroding public confidence in the profession's fiduciary commitment, this attitude is "shocked" into the realization that what we once believed to be "real," or "true," can be otherwise. The profession is attempting to maintain or reestablish such an epoché so that the patient will accept its power and authority. The shock to the taken-for-granted trust in physicians is reconciled or explained away, and the epoché is restored. In this condition, trust is used to reinforce a fixed means-ends orientation which obscures the value-judgments held to be meaningful by the medical profession's members. "It allows reality to be disclosed only in accordance within certain predetermined possibilities, yet it represses the fact that things have been interpreted in this manner" ([58], p. 670). This order of trust is termed "trust-as-control."

It is ironic that the profession places such an emphasis on trust and trustworthiness. Hawthorn argues that "aristocracies" (or privileged social institutions) create trust by creating distrust, i.e., the medical profession creates trust in itself by encouraging public distrust in other, competing experts [125]. The turn from homeopathic to allopathic medicine early in this century was due, in part, to the success of the American Medical Association in discounting homeopathy [262]. Ironically, one reason for homeopathy's recent popularity is its focus on the individual. Homeopathic practitioners traditionally spend considerable time, commonly an hour long consultation on a person's complaints and general health, finding a unique course of treatment for each person. This is in direct and appealing contrast to the medical assembly lines many patients encounter currently [102]. Aside from the debate about the efficacy of homeopathic medicine, this emphasis on the personal touch raises a second level of trust.

The second dimension of trust is a deeper, more complete condition that has similarities to an act of freely given and received faith that transforms the person's relationships with others. It rests on an opinion or decision based on past experience, demonstrated skill, and evidence of benevolence. This trust involves a calculated risk based on evidence that is still not conclusive enough to compel assent or guarantee success – a trust "in spite of" the uncertainty and incompleteness of human finitude. The second instance of trust results from the reintegration of a more comprehensive understanding of trust, and makes possible an authentic, mutual decision to be trustfully responsive and present to each other in relationship. Attention to the virtue of "trust-as-faith" suggests a way to reinterpret the fiduciary component of medical practice. Since the one who gives and the one who receives trust in the medical relationship are linked and mutually dependent, the recovery of a theological view of trust-as-faith represents an appreciation of interdependence in medical matters.

WHY IS TRUST A PROBLEM?

The unequal balance of power between physicians and the public in terms of knowledge and the development of professional standards clearly places the public at a disadvantage [223]. Unable to impose ordinary economic controls upon the monopoly of the profession, society has developed an emphasis on the essential ingredient of trustworthiness in relations between doctor and patient.[8] As the profession's knowledge, expertise, and, consequently, power grow, people begin to demand more exemplary performance, more certainty, and more guarantees as prerequisites for their trust and confidence. If trust begins to lose its efficacy as a form of social control, a second manifestation of trust as a form of social control is revealed: litigation, regulation, and accountability increasingly are emphasized as the means for control of the profession's power ([8], p. 22).

Public attitudes towards the medical profession have always been ambivalent, and these control mechanisms have a long history ([77], p. 515; [219]). However, recent manifestations have a sharp edge both in their sociological and psychological dimensions. Public ambivalence is caused by the structured expectations doctor and patient have of each other, as well as by physician deviance from expectations or malpractice [188]. Distrust is caused by an individual's failure to embody traits associated with a profession, or the failure of a profession to inculcate such traits. Distrust also results from a professional's pursuit of self-interest. For example, in the name of acting in the patient's best interests, a doctor may act to preserve the profession's authority. Yet this action can represent self-interest in that it serves to maintain the physician's own power (in an act of paternalism) ([149], p. 97). Professional altruism, so powerfully influenced by an ethical obligation to benefit the patient (an obligation that doctors and not necessarily patients have imposed on themselves), can cruelly deceive both the patient and the physician. A fear of abuse and deception tempers unconditional trust.

As a result of the social organization of medicine, the profession guards its independence and autonomy against what it terms "lay control" ([87], pp. 110ff; [216], p. 470).[9] As the profession becomes specialized, as the scientific method is used to generate powerful knowledge and expertise, and as concern for the high cost of health care mounts, the public's uneasy ambivalence finds new strength in a call for health care reform. The profession naturally acts to reinforce its authority and maintain its control by emphasizing its area of primary strength – expert knowledge and decisional autonomy, shared with colleagues and aimed at patients' benefit.[10] Polarization between public and professional interests places more strain on the correlative fiduciary aspect of the relationship and leads to a climate of suspicion in which the assumption that the medical profession is acting for the best interests of clients is undermined. An economy of trust is established in which trust becomes a commodity to be bartered, and the profession's fiduciary commitment to the well-being of the client becomes an illusion of trust.

Since World War Two the appropriate use and limits of the medical profession's power and authority have concerned many people within and without the profession.[11] This discussion of limits has resulted in practical benefits for clinical care. During this same period, however, a climate has developed in which many people seeking help from physicians approach relationship with them more cautiously. In addition to government involvement in medical matters, legal restrictions on the actions of physicians, and an emphasis on auditing and monitoring physician performance, distrust serves as a "functional equivalent" of trust intended to promote effective social control of the medical profession. These developments affect the fiduciary component of the medical relationship. Many doctors deplore this distrust, claiming that it represents an attitude of anti-science and an attempt to undermine physicians' authority and autonomy. Such distrust, they claim, hampers doctors' efforts to do the work to which they are dedicated – to fully serve in the best interests of their patients. I argue that the public's use of distrust as a means of social control is nurtured by the profession's commitment to medical science. Due to the profession's zealous dedication to the cultivation and utilization of knowledge, ironically the expectation of confidence in the physician's beneficent trustworthiness is disrupted.

A decline in trust is not absolute but relative to changing expectations. It may be that the expectations leading the public to grant trust to the medical profession are changing. Such a shift in perception or expectation is not unusual. A given group's knowledge, self-control, or responsibility may be increasing or decreasing over time, making that group more or less professional ([8], p. 136). The growth of scientific medicine has created new expectations of physicians. At first, the expectation suggested that a scientific basis for medicine would lead to more knowledge, more skill, and more technological prowess. This increase would lead, in turn, to better outcomes in health care situations.

However, the increased expectations have not been met (see [51]). As the medical profession has become more powerful in terms of formal knowledge and technical expertise, the public fears that the fiduciary commitment that characterizes the medical profession's claim to "status" is eroding. It is one thing to trust physicians' promise to diagnose, treat, and not to make matters worse unnecessarily. It is another to trust them to know what is beneficial to their patient, in the many senses of this word, when choices are available which can make matters both better and worse ([149], p. 94). Answers that satisfy everyone's varied concerns cannot be reached on the basis of medical judgment alone. Science increases options, but each option entails its own significant benefit or risk. The public may seek to insure its own protection by limiting professional power through increasing its own autonomy and control.

I do not mean to claim axiomatically that all doctors are suspicious of all patients, or that all patients mistrust all doctors. Lionel Trilling believed that there is a modern tendency "to locate evil in social systems rather than in persons" ([59], p. 26). One of the first things that should be said is that many physicians are aware of the limitations of their enterprise, and are concerned about the dangers of false claims to certitude in their work. To a real degree, the changes in medicine over the last twenty years are due to the willingness on the part of many physicians to question the presuppositions that underlie biomedicine and the institutional structure of their profession and health care generally. However, I will argue that many patients are growing distrustful of physicians, that increasingly society does not take for granted the profession's traditional claims to authority and control in its work.

In the current climate the danger is a professional retrenchment in structures of power, even as "reforms" are made (for example, attention to patient autonomy; the acceptance of living wills; or concern with cost containment). Many of the issues in medicine appear to be issues of control (for instance, control over dying or end of life decisions, over authority in the work place, or the broader focus on cost containment and health care reform). This concern with control leads to shallow, adversarial relationships if the struggle denies the necessity to trust others. Because the physician-patient relationship is one that inevitably involves structures of trust, loyalty, and power (due to the disparity between a vulnerable, needy person and one whose help is sought), it is no wonder that trust is a concern for patients and physicians in late twentieth century American medicine.

THE PROBLEM OF TRUST IN THE POSTMODERN AGE

It is not coincidental that the sense of uneasiness in the patient-physician relationship corresponds with the effect of postmodernism. In this discussion, *postmodern* refers to a period of time, the present and future, marked by a particular style of criticism. The postmodern period is a time that follows the atrophy of modernity's various projects in science, ethics, aesthetics, and theology, and declares the collapse of the rationalistic aspirations of the Enlightenment ([284], p. 115). Postmodernism is a style of cultural analysis that seeks to reveal the cultural construction of concepts people generally assume to be natural or universal.[12] Proponents of this postmodernism seek to deconstruct such concepts as God, science, medicine, principles such as autonomy or justice, and so forth, to break the grip of their control on our thoughts and actions, revealing them to be cultural constructions of meaning, socially produced, grounded in issues of power and control. With regard to medicine, the postmodern period is characterized by the increasing effectiveness of medical technology, but also its increasing cost and, in response to that cost, increased government intervention and growing public awareness of scarce resources for care. If there are no foundational truths to serve as the basis of our relations with one another, where once one could assume that the image of the physician represented trust, predictability, and accountability, we would seem to be left merely with relationships of mutual self-defense and self-interest. It is no wonder that contract language is common in an effort to ensure these characteristics.

Since the time when the term "metaphysics" was first applied to the text that contained Aristotle's investigation of the essential nature of Being qua being, the term has been used to refer to such an investigation ([238], p. 5). In Western metaphysics, the study of Being *qua* being (ontology) is tied to theology, the investigation of the highest being which is the necessary condition for the possibility and actuality of all other beings. As a result, in deconstructionist eyes Western philosophy and theology have become "logocentric," committed to a belief in some ultimate "word," presence, essence, truth, or reality that will act as the foundation of all our thought, language, and experience. According to Derrida, Western philosophy seeks the sign that will give meaning to all others – the "transcendental signifier" – and to the unquestionable meaning to which all our signs can be seen to point (the "transcendental signified") ([56], p. 49). In theology, "God" is that sign, and attempts to develop a covenantal basis for social relations grounded in this sign become, for deconstructionists, logocentric fictions characteristic of the modern post-Enlightenment age.

To put this another way, postmodern critics attempt to subvert what they find to be the tyranny of comprehensive worldviews or visions of the *whole*, in relationship to which *parts* derive their meaning. Such visions of the whole are characterized by Jean-Francois Lyotard as "metanarratives" peculiar to modernity and its distinct aspirations ([171], p. 84). The idea of health as a condition to which we have access with the right biomedical knowledge and understanding represents such a modern metanarrative. The concept of "health care," in relation to which all of the institutions, models, roles, and values associated with that sign take their meaning, reflects a complex of relationships intended to correspond to this modern "reality" of health. Within this metanarrative, relationships are hierarchically arranged, inherent *power-plays* that exclude, distort, and suppress any element that "threatens" them ([284], p. 114). The physician-patient relationship is an example of such a power-play.

Physicians' appeals to biomedical authority reflect a narrative of mastery over natural, biophysiological processes; for example, because of their power-full knowledge, physicians are able to act as gatekeepers to health. The professionalization of their knowledge-based skill becomes a power-play that subordinates patients in clinical relationships.

By drawing attention to the ways in which overarching concepts are actually culturally constructed and are not the universal truisms that most people assume, deconstructive postmodernism raises the question of the illusion of trust in the medical profession's traditional claims of service. Rather than assume that physicians use their superior knowledge and technical expertise in the best interests of their patients, in the postmodern atmosphere it is easy to critique this fiduciary commitment as arbitrary, the product of one group's self-serving grasp for social power. Rejecting all metanarratives, or supposedly universal representations of reality, deconstructive postmodernists insist that the meaning of every aspect of human existence is culturally created and determined in particular, localized circumstances about which no generalizations can be made. All knowledge, then, is situated within a culture.

The deconstructive stance undercuts the idealized "truth" of the Enlightenment that the autonomous man of reason is in control of his possibilities or choices ([261], p. 14). This analysis emphasizes the partial truth that a human is born into a set of cultural constructions and constraints, and lives out his or her life in the embrace of various "discourses," socially invented systems of perception, meaning, and knowledge. "I" am not autonomous apart from this embrace. There is no autonomy; one can only act self-consciously "as if" one's actions or responses had meaning. Any claim for a professional fiduciary commitment is merely socially produced discourse composed of self-referential words and concepts. Such a claim serves only as a model to act as if one were engaged in service, but the only real commitment is to the ideology of atomized detachment ([261], p. 16).

In sum, this postmodern style of criticism attacks any thought-system that depends on a first principle or unimpeachable ground upon which a whole hierarchy of meanings is constructed. The quest for the "thing in itself" is based on a misapprehension: no final meaning can be fixed or decided upon; thus, interpretation abounds. This attempt to expose the illusion of metaphysics (that behind, beneath, or within the play of appearances there is a truth which can be discovered, uncovered, or disclosed) serves to undercut the pretensions of secularization, as well as theology ([284], p. 114).

It is not surprising that the postmodern style is tinged by an undercurrent of suspicion: claims to univocal meaning, final truth, or logos are continually undermined, and we are left with the finitude of perspectivism. In this environment, any tradition that was formerly taken for granted is undermined, including the medical profession's traditional fiduciary claims. In short, there is a serious challenge to trust in the clinical context, what Pellegrino calls "an ethics of distrust," i.e., the formal attack on ". . . the very concept and possibility of trust relationships with professionals" ([224], p. 79). Such an ethic places tighter restrictions on physicians or eliminates the need for trust entirely. In the postmodern age of suspicion, trust cannot be assumed or taken for granted in our relations with professionals.[13] Although trust always has been fragile in social relations, people now seem more conscious of the risks of trusting others. As a result, both physicians and patients are seeking protection and control, to reduce or eliminate the disruption in their lives represented by their respective encounters with illness and with each other.

The deconstructive stance just described discards the very concept of truth itself as metaphysical baggage. Such *strong* postmodernism regards all totalizing visions as finally ontotheological illusions, seeking to maintain a hierarchy of privilege by obliterating heterogeneity. Instead, the strong postmodernist position allows no shred of essence to remain untouched. There exists only "sheer heteronomy and the emergence of random and unrelated subsystems of all kinds" ([109], p. 41). As a result, strong postmodernism borders closely on nihilism.

A *moderate* postmodern position takes seriously fundamental postmodern concerns such as the radicalness of historicity, the pervasiveness of ideology, the decentered subject, the rejection of transcendentalism, and the too-often, too strict bounds of rationality. Moderate postmodernism, however, seeks to defend a rationality that is appropriate to the newly "presenced" postmodern perception that foundationalism is always elusively absent and that attempts to establish "final" or universal truths appear condemned to failure. Rationality in this view must be "consistent with our finitude, with our historicity, with the dependence of thought on changing social conditions" ([109], p. 42). Reason is exercised in circumstances which are thoroughly finite, conditioned, and historical, yet the moderate position recognizes that presence and absence are commingled (both hidden and revealed).[14] As a result, the moderate postmodern stance allows a nonfoundationalist phenomenological and sociological analysis to uncover the historical and hermeneutical dimensions that inform our efforts to make sense of our existence.

THE PHENOMENOLOGICAL ANALYSIS: WHAT'S GOING ON?

Like a serpent who lives in both water and land, the phenomenological method provides a way to cross boundaries. In the phenomenological attitude, the interpreter leaves his or her own familiar world for a moment, crosses into the unfamiliar world of the "other," and returns with a knowledge made possible by the crossing. Often, he or she appears to be in two worlds at once, striving to maintain a dialogical relation between self-understanding and understanding the other. Thus, work in the tradition is reflexive. For a study of what trust means for and to the medical profession – trust which itself crosses the boundary between the known and the unknown – this attitude is appropriate for the task ([24], p. 1).

In its philosophical origins, particularly in the work of Edmund Husserl, phenomenology intends to uncover the essence of a particular phenomenon. While the range of Husserl's work extends beyond the scope of this project, it is important to note that Husserl's phenomenology was a method and an attitude, a way of critically approaching the world in an effort to gain insight into it ([138], p. 19). Husserl designed his method of investigation to reconstitute the appearance of the world in terms of its transcendental structure or essence.[15] From the perspective of the person's ordinary or typical attitude, he believed, this essential dimension is hidden. In the phenomenological attitude, in eidetic analysis, attention is shifted away from the object perceived (its facticity or noesis) to the perceiving itself (the essence, or noema).[16] In this reduction or epoché, Husserl believed that it was possible to bracket those views relating to the historical stance of the natural attitude that cause one to base his or her conclusions on historical facts and preconceived values concerning those facts. The *natural attitude* is the term used to encapsulate the essential presuppositions, structure, and significations of the world of everyday life. It is the standpoint that reflects an individual's ordinary, unquestioning outlook on daily life in which the person assumes that reality corresponds to his or her understanding of it. In phenomenological theory, one can become deliberately, systematically disengaged from the natural attitude through the method of the epoché or "suspension of judgement." The epoché "brackets" or suspends the taken-for-granted understanding of reality. Husserl gave the terms their classical formulation (see [197], pp. 34–44). By freeing one from a fixed subject-object dichotomy, this form of radical criticism presents a more complete understanding of social action and meaning. Freed from these "blinding" preconceptions and typicalities, phenomenology enables one to re-view experience in the world.

However, postmodernism challenges Husserlian claims for the ontological status of the self as the transcendental ego. According to postmodernism, there is no such ego. One way to bypass the issue is to imply an ontological primacy for social and cultural reality over the reality of selfhood; this position involves the philosophical assumption that the self comes into being only through interplay with the outer world. Thus, ontological reality lies in multiple levels of reality; there is no transcendental ego, no universal position from which to understand and to act. Who we are, and how we can know another, is always somehow hidden yet revealed, in the "unending play of surfaces" ([267], p. 16).

While the postmodern critique suggests that the attempt to reveal what is hidden is neverending (a significant theological observation), many of the insights of phenomenology have enriched the work of various disciplines. In the sociology of knowledge, for example, phenomenological categories and method are utilized to pierce the veil of appearances and uncover the basic meaning structures within which individuals make sense of their social world ([243], p. 208). The phenomenological attitude and method enables the investigator to suspend the conceptualizations he or she brings to the phenomenon and opens the way to "see" something of what is "going on." Given these presuppositions, it is possible to present a phenomenological and sociological examination of the way in which the natural attitude constructs and is supported by the system of relevances with which a person makes sense of the world. Thus, the goal of phenomenological sociology is not to philosophize about the nature of existence or to refine a metaphysics of being, but to interpret these systems of relevances on their own terms.

Phenomenological sociologists would argue that what it means to be a physician is constituted by the definitional boundaries of the social institution of which he or she is a member. These boundaries are determined internally by a coherent and complex system of knowledge, skills, standards, values, and organization. The phenomenological "perspective" of the physician differs from that of the patient, a difference which must be taken into account in physicians' moral considerations and decision-making.[17] For instance, Wendy Carlton describes a "clinical perspective" that characterizes the approach of physicians to their work and their patients, one which has an effect upon moral decision-making ([37], pp. 65–83). This perspective allows physicians to "set aside" the expectations of everyday reality and engage in a medical "habit of mind" which determines their understanding of the situation. For instance, with the biomedical model the physician works within an "accent" or "habit of mind" which primarily is based in the "finite province of meaning" of medicine.[18] It is this habit of mind which marks the medical profession as a particular institution within society. As an institution, medicine is an amalgam of conceptual and social organizations with an historical tradition, the combination of which determine its definitional boundaries. Such boundaries also are determined externally by the profession's complex matrix of relations with other individuals and institutions in the world of working. If the public's trust in the profession, and by extension in its members, is eroding, there will be a concomitant effect on the way in which the profession and its members construe the meaning of trust – in themselves, in their work, and in their patients. In this perspective the physician sees his or her own trustworthiness as a taken-for-granted ingredient in being a doctor.

The present relationship between physicians and the public indicates that physicians' thought and practice, on the one hand, and the beliefs and expectations of their fellow citizens, on the other, are uncoordinated and disharmonious [277].[19] Therefore, a phenomenological "reduction" of the medical accent of meaning invites a re-interpretation of trust's importance in shaping the boundaries of professional identity. This interpretation reveals the roots of the uneasiness and resentment that complicate relations between physicians and patients, and suggests possible directions for a recovery of mutual trust.

Physicians complain that ethicists too often leave the situations of everyday medical work to the doctors while developing "compelling arguments" drawn from moral principles, which they then apply to the physician's work. However, while physicians recognize the significance of bioethics' "landmark cases" (the ones with which bioethicists seem most interested), the moral issues that trouble physicians are more usually the ones related to everyday concerns: relations with colleagues, interactions with patients, and the activities of everyday practice. A phenomenological approach to medical "reality" stresses that moral issues must be presented solely within the context of their actual occurrence – moral analysis must begin at the bedside, with an appreciation and understanding of medical "experience" ([295], p. 27). Therefore, the application of a phenomenological analysis of medicine must precede any discussion of moral principles or theological analysis "from the outside in."

If theological ethics is to be taken seriously by physicians, the contribution of a theological morality to medicine must be demonstrated. Talk of trust becomes meaningful only in situations of uncertainty and risk, yet trust does not eliminate the risk – it becomes meaningful in the face of risk. Although medical work and physicians need not be religious to be efficacious, theology can contribute to an understanding of these situations in which risk and uncertainty are present.[20] The theological themes of alienation and reconciliation suggest loss and recovery in such boundary, definitional, or limiting situations.[21] Trust represents a "leap" of faith that occurs in those situations and moments of uncertainty and risk in which assumptions regarding meaningfulness of life are challenged. Religious faith provides a confidence and trust in the face of these "limiting" moments.[22] In faith, the self is transformed and meaning is (re)established.[23] Again, the elements of risk and uncertainty are not eliminated; the possibility of meaninglessness remains. However, a transformation of faith can provide meaning in human life and restore a more integrated "natural attitude" to one's experience of the everyday world.

An erosion of trust exists that is affecting the medical profession's sense of its traditional fiducial commitment to its clients. The alliance of medicine with science, coupled with social changes, has made the profession and public more self-conscious of the asymmetrical nature of their relationship. With the perception of cultural pluralism and the loss of an "assumed" moral consensus, participants in the medical relationship are becoming uneasily aware of the constituted nature of trust. A more controlled and self-possessive approach to medical encounters is developing on the part of patients towards physicians, but also on the part of physicians towards patients. Under this condition, the fiduciary dimension of medicine takes on an illusory quality.

FROM PHENOMENOLOGY TO THEOLOGY IN THE POSTMODERN AGE

Post-Enlightenment criticism of a theological basis for ethics has created a search for final, universal moral precepts and principles to which all rational agents can pledge allegiance. With the seemingly insurmountable problems of subjectivism and relativism, however, it appears that this search has negated itself in the Western, postmodern world [32, 234]. Moreover, this critique overlooks the possibility that faith can inform reason, not in the sense of claiming that only the person who has been transformed by faith will "see" rightly, but that theological discourse can broaden understanding of medicine's professional ethics.

Theology owes a debt to deconstructive postmodernism for bracketing the naivete of the constraints of modern rationalism and scientific empiricism, exposing the dogmatism of the socially constructed relation of metanarratives and discourse. However, theology in the postmodern age recognizes that all beings are internally constituted by their relations to others. We are not the fixed, self-contained entities of the modern world. Instead, we are ever changing and ever interconnected with other humans and the natural world. Our cultural interpretations of reality are impoverished if they operate in isolation from this larger, interconnected context ([261], p. 20). A postmodern, self-critical theology has much to offer as we move beyond a self-directed, self-protective relationship of physician and patient.

Theology from time to time must reconsider its own interpretative schemes, how it is to understand the divine-human relationship, and how it thereby constructs a world of meaning. Theological worldviews can become taken for granted and run the risk of becoming ideologically bound. The encounter of theology with postmodern concerns can "shock" the theological natural attitude into a reworking of its taken-for-granted conceptualizations. In the postmodern culture of criticism, in considering issues of faith and trust in medical matters, this issue becomes important. How is the term "God" being used in a discussion of the role of a fiduciary commitment in defining the boundaries of medical authority?[24]

A Traditional Theological Basis for Ethics

According to the tenets of classical theism, God is the supreme creator, who brings the world into being and directs its course. This First Cause (*Archē*) is also the Final Goal (*Telos*) of the world. Utterly transcendent and thoroughly eternal, God is represented as totally present to Himself (sic). He is the omnipresent source, ground, and uncaused cause of presence itself ([267], p. 7). The term "God" as used in Western culture stands for or names the ultimate reference point or orientation for all life, action, and reflection.[25] The self, it is believed, is made in the image of God and, consequently, is also a centered individual, both self-conscious and freely active. Together, self-consciousness and freedom entail individual responsibility. History is the realm in which divine guidance and individual responsibility meet. It is believed that history is not dominated by individual randomness, but begins with a definite event and extends through an identifiable middle to an expected end. History becomes a teleological process whose meaning can be coherently represented ([267], p. 7]. God, self, and history are thus bound in mutual and interactive relationship, one to the others.

It is the solid, common understanding of this interrelationship, this "reality," that is threatened in a culture of criticism. For strong postmodernism, God as other is a characterization, a metaphysical illusion. I maintain that in the moderate postmodern culture of criticism, the depiction of God as an entity, albeit exalted and supreme, remains a common characterization for people. Yet a suspicion that the scientific, ethical, and theological under-pinnings of modern life are collapsing serves to call into question understandings that once served as the basis for our relations with one another. As a result, people seek trust (as control) in that which they can see and control. They seek to make the world "familiar" and dependable.

This crisis focuses our attention on the heart of the theological purpose. From the theological perspective, the understanding of God orients human life. However, while a religious symbol gains its specific content from particular cultural traditions, it stands for more meaning than it can convey. As a limiting idea of reality, therefore, an understanding of God can be approached but never literally grasped. Human understanding of God is a historically conditioned construct, subject to continual criticism and reconstruction in light of changing perceptions and understandings over time [210].[26] The task of theology is the investigation, elucidation, and exposition of the historical center of the faith of the Christian community ([151], p. 109). As such, theology raises ontological issues because it does not dwell completely within the pre-interpreted "givenness" of things as ontic. It seeks to call into question the normative conceptions which are taken-for-granted in the natural attitude. In doing so, it opens the possibility for critical reflection upon, and transformation of, the way in which meaning is construed ([58], p. 672). Theology becomes the "wound that heals" in a double movement of uncovering what is present, although hidden, and disclosing what is absent, yet revealed.

In a theological review of the definitional or "limiting" meaning of trust for physicians, an examination of the importance of knowledge and trust in shaping the fore-knowledge with which physicians approach their world focuses on the "moment" in which familiar reality is called into question. Medicine represents a process, a continuous series of relational moments, in which the content of knowledge must undergo occasional revision. This moment becomes the starting point for a medical theological ethics which seeks to recover human relationship in trust-as-faith.

However, Christian theological ethics exists in the postmodern world with contrasting, competing explanations for meaning in human existence. Theology is one among many systems which offer relative explanations, each claiming authority. Under the pervasive climate of suspicion that characterizes the modern world, the conception of God as transcendent and immanent is subject to renewed debate. For many people, this tension is showing signs of fracture, leading to a loss of trust in the first conceptualization of God and a quest to secure the latter. This movement pushes for control and domination which obscures an appreciation of God as "other," and relativizes or trivializes God as immanent.

Theological Concerns Raised by Trust-as-Control

A theological analysis highlights the use of trust-as-control by humans in relationship. A theological discussion begins with the declaration that separation from God creates the basic human condition of dependency from which the awareness of incompleteness and neediness flows. By bracketing the traditional theological model of the transcendent, wholly other God, the dilemma of relationship in the post-modern world is revealed. In a culture of criticism, the recognition of multiple levels of reality raises the conflict of relativism and pluralism. The perception of multiple competing groups also creates an internal awareness of the self-as-other ([209], p. 17).

A crisis of authority at all levels of social life is the result. Faced with the inescapable condition of dependency, people seek meaning in their existence which is dependable, which they can take for granted. The difficulty of bracketing the shocks to the natural attitude that exist in a pluralistic world undermine people's certain confidence in values. We may deny the conditions of uncertainty and neediness because they make us uncomfortable. To deny our uneasiness, we focus on and demand what we desire because this demand implies more control in the face of our basic dependency. However, this desire for control is itself a reflection of dependency. It is a more aggressive, self-directed effort to cover up neediness, but the satisfaction of what we want or desire cannot be satisfied. Dependency cannot be eliminated from our lives.

In the effort to eliminate the neediness flowing from the human condition of dependency, both physician and patient retreat from authentic presence in the health care relationship so that the other can fill the gap – and prove his or her trustworthiness. However, this movement creates a giving and receiving that is self-directed and intended to control the presence of the other. The struggle for control affects the fiduciary component of the postmodern health care relationship. Instead of giving (and receiving) trust offered by the other's presence, thereby accepting responsibility for one's actions and responding to the other's actions, each party struggles to eliminate risk by controlling the other who has become a threat to the self's image.

A recurring theme in Western theology is the effort of humans to achieve a position of dominance in relations with others. This effort appears to grow out of the conviction that mastery results from the ability to secure presence and establish identity by overcoming absence and repressing difference.[27] Similarly, in matters of health care, given the inevitable diversity of individual experiences and religious and moral conscience, conformity to a single moral norm cannot be attained without the threat of, or the fact of, the repression and persecution endemic to paternalism and coercion.[28] The struggle for mastery, however, is always self-defeating. For theists or atheists, the result is shock of confrontation with otherness that can no longer be bracketed. We are driven to minimize through domination or submission the risk of relationship. More basically, this drive is based on the denial of the human need for relationship and human interdependence. Theological analysis of the drive for domination offers a counterpoint and a new possibility for relationship.

A MAP OF THE METHOD

The practice of medicine is not a theoretical and value-free science, but an "hermeneutical" enterprise. Hermeneutics refers to the study or science of the act of interpretation. With roots in ancient Greek philosophy, hermeneutics has come into prominence again over the last two hundred years in the humanities and social sciences ([213], pp. 3–45). Its application to medicine is clear, because the diagnostic situation is always an interpretive one. An investigation into the use of trust-as-control also requires an interpretation of the tension of perspectives which arise in the practice of medicine.

The physician is an interpreter who is constantly trying to understand data in the context of the particular clinical situation. A certain set of findings are "read" in the hope of uncovering the organic condition that constitutes its "meaning." Moreover, interpretation is always closely tied to application. The physician's best reading of the situation determines how diagnosis and treatment will proceed. Diagnosis and treatment are always open to alteration or refinement as a result of new developments. The physician-hermeneut employs a set of interpretive tools learned in school and subsequent practice for unravelling the "text" or a "living human document," that is, the actual patient at hand ([53].[29] However, Zaner observes that symptoms are uniquely the patient's own, their meaning being "textured by that person's own biographical situation." But modern medicine "displaces" the patient's "situation" and sees his or her symptoms as only significant in the physician's terms, according to the biomedical model ([11], pp. 41–42).

This displacement has contributed to the erosion of trust in relations between the public and the profession. Each participant in the relationship is subject to interpretive modes that shape human existence on individual and social levels. Each has a pre-conceived conception of what the physician "should" be like and of what trust means in the medical relationship. These covert value-judgments are present in the way people "see" and interact with the others they encounter. If the other does not match a person's expectations and cannot be incorporated into his or her interpretive schema, distrust or trust-as-control is utilized to ensure a sense of "reality." The physician/patient "sees" the patient/physician from within a horizon of meaning and seeks to fit the other into the anticipated frame of reference. To understand this dynamic, we require a hermeneutical approach which entails a double movement of regression and progression: a disclosure of or "making present" of the "hidden" structure of meaning. Thus, determining the situation requires a hermeneutics of daily life, or to use Gibson Winter's term, a phenomenology of "the world of every day" ([289], p. 154). This re-interpretation brings new possibilities to light.

The Need for a Hermeneutic

Hermeneutics has importance for a study of the fiduciary aspect of professional identity. The physician must interpret the meaning of being a physician within the world of the medical tradition and within the everyday world of working. He or she is part of the medical profession's ongoing tradition. This tradition constitutes his or her understanding of the meaning and value of medical work. At the same time, the profession is part of the totality of human culture and represents a "closed world" of structured meaning, the accent of which shapes the doctor's self and his or her relations with others. Finally, the physician also is involved with the "living human document" or "text" of the patient. The physician's pre-understanding of medical work (including the doctor's understanding of knowledge and fiduciary responsibility) governs his or her response to the patient who also possesses a particular pre-understanding of the nature of the physician and medical work. The encounter of these "worlds" raises the issue of trust both as the confident investment of meaning in one's understanding, but also as the possibility of encountering new meaning in the risk of relationship.

The elusive quest for a metaethical or foundational basis for medical ethics is not my concern. Phenomenology was developed to establish pure observation as a foundation for empirical science, but it is now recognized that there is no neutral stance, no place from which to make observations that is not *a place*. Thus, phenomenological work has led to an appreciation of the social constructivism of our life-world ([268], pp. 1–11). From this point of view, diagnoses, diseases, medical knowledge, and health care institutions are considered social constructions, which can be understood in their empirical social and cultural context. The emphasis is on the constructs that comprise medical *practice*.

Rather than seeking a way to secure a final understanding of the fiduciary nature of the patient-physician relationship, we need to develop a hermeneutic that will place the review betwixt and between taken-for-granted assumptions regarding its nature. The hermeneutical trajectory of foundationalism (the Husserlian stance of objective interpretation) claims that it is possible to recreate the sociocultural world of the text and, within context, to determine its original meaning. However, understanding is always interpretation, and understanding is an event over which the interpreting subject does not ultimately preside ([109], p. 38).

In order to develop this hermeneutics of the life-world, I follow a two-step process. First, as preparation for the second part of the volume, chapters two, three, and four describe the definitional boundaries of the medical profession from a sociological and phenomenological perspective. A social model of the medical profession can be established by the discussion of finite provinces of meaning within the construct of social reality [243].[30] The profession can be described as a finite province of meaning, a "practice," possessing a certain "accent of reality" through which it views itself and the world.[31] In particular, I will argue that the physician as a member of a profession encounters his or her work from within a "horizon of meaning" – what Heidegger refers to as a "fore-structure of understanding" ([275], p. 220; [127], pp. 188–195). This horizon of reality is constituted by a tradition, partly institutionalized and partly informed by an individual's biographical situation. It represents a certain way of understanding, determining, and acting in and on the world, as well as certain habits of mind and relevances, all of which give meaning to the work of the physician and give to him or her a sense of identity and authority in the world of everyday life. The profession as a community is structured around a central focus and a commitment to other members of the profession. The community's horizon of meaning shapes the perspective of its members regarding the purpose and value of their work. It also shapes the social relationships that its members have with the world of everyday life. Their accent of meaning bestows a sense of "immediacy" or "reality" to their province.

For the most part, the contemporary physician tends to understand the condition of a patient through a single interpretive paradigm – the biophysical model. The roots of this model rest in the anatomical-physiological understanding of the human body as a causally-determined machine [74]. When, for example, a pulmonary specialist is called in on a case, he or she comes armed with a set of hypotheses, techniques, and principles that orient the physician's interpretation of the situation, even if these are never made explicit. For the pulmonary specialist, this horizon of meaning includes general cultural assumptions concerning the nature of the material world and the human body, as well as more specific knowledge concerning the physiology of the lung, its relation to others organs, and the lesions and modes of decompensation to which it is susceptible. While a given case may be ambiguous, the conceptual universe within which the pulmonary specialist operates seems to form a coherent, self-consistent, and unified structure.

Within the profession's horizon of meaning, there also exists a fundamental understanding of the characteristics of the doctor-patient relationship and the appropriate roles each party is to play. The physician sees himself or herself as the primary, if not final, authority in the care of his or her patient. The physician tends to resist criticism of this interpretation of the relationship, arguing on the basis of his or her superior knowledge, technical competency, and the profession's sense of its traditions and values. Criticisms of the profession's interpretation often are perceived as threats to the authority that enables the doctor to do his or her work and do it well (in the best interests of the patient and public).[32] An eidetic interpretation of this model of the medical profession will support the claim that a climate of distrust has developed between the "technically competent" physician and the patient that undermines the fiduciary dimension of the doctor-patient relationship.

A review of the phenomenological dimensions of the medical profession as a finite province of meaning within the everyday world demonstrates the importance of the acquisition and control of knowledge in the "reality" in which the physician operates. Although in the 1990s the physician's field of knowledge extends beyond a mere commitment to the biomedical model of illness, this model still remains fundamental to the "practice" of medicine. The particular nature of the model shapes the profession's accent of meaning, its clinical vision, within which the physician approaches his or her work and relations with the public. The analysis of the role of knowledge in shaping the profession's accent of meaning will introduce the importance of trust in shaping professional identity.

I do not argue that trust as a representation of fidelity in the medical relationship is the preeminent virtue, nor is it the only virtue important to interactions between doctors and patients. However, the importance of community and tradition to moral reasoning and conduct, and the role played by the virtue of trust in this on-going process, offers a context for a theological reconstruction of the medical profession's accent of meaning and the effect on the physician's fiducial commitment of using trust as control. An exploration of the role of trust in the physician's accent of meaning and in shaping the physician's interpretation of the experience of medicine opens a conversation with theology.

The Theological Exploration

Having set up a model of the profession as a finite province of meaning within a social matrix of multiple levels of reality, chapters five and six will be a theological exploration of the dynamics of giving and receiving exhibited in the primary health care relationship between physician and patient. I will argue that the present model of the relationship, grounded in the biomedical model of illness and the climate of distrust, results in a relationship in which the giving and receiving of trust are brokered by self-interested, self-seeking adversaries alienated from each other and from themselves. A theological inquiry contributes to a reinterpretation and appreciation of trust-as-faith in physicians' work.

A phenomenological "bracketing" of the taken-for-granted accent of meaning that characterizes the medical profession raises a theological focus. As human beings, what we ordinarily assume to be meaningful in our lives largely is constituted by the horizon of our particular province of meaning. However, a phenomenological analysis of the nature of social reality reveals any province to be an artificial construct. On occasion we are shocked to discover the "facticity" of our natural attitude – that the world can be other than it appears to us. It is a peculiar feature of human nature to desire to extend this facticity beyond the limits of our province, to deny our essential dependence upon what is beyond our control, to deny our social existentialism (see [201], pp. 234–256).[33] Inasmuch as the arena of health care, and primarily the physician-patient relationship, deals with issues of life and death, well-being and dis-ease, it reflects this desire to control our human neediness. Health care becomes an effort to satisfy our desire to deny our human condition.

Due in part to the promise of biomedical science to cure our ills (our needs) if given more money, more support, and more trust, the technical knowledge and skill of the physician increases. The promise of a "remedy" for human neediness, of course, remains unfulfilled because it cannot be satisfied.

A theologically descriptive task that begins with a phenomenological bracketing of the natural attitude of everyday life (even the natural attitude of groups within the Christian community) reveals the roots of the fragmentation of moral experience. The natural attitude generally takes for granted particular assumptions and expectations of perceived reality. In the postmodern culture, people are aware of "otherness" (that things could be other than they are believed to be), yet struggle to maintain their assumed view of reality against its encroachment. As a means to retain trust and social interaction, such a morality becomes grounded in an appeal to rights language which at best insures a condition of mutual non-harm.[34] However, I would argue that attempts to maximize individual autonomy are more indicative of an erosion of trust among members of the social matrix. As agents become possessive of their assumed autonomy, they become caught in an illusion of self-control. The roots of trust-as-control lie in a denial of the basic condition of dependency and neediness that motivates a drive for self-control, even as it alienates the members of an intimate social relationship such as the one between physician and patient.

A phenomenological and theological investigation of the medical profession's fiduciary focus suggests a manner by which trust's meaning in medical matters can be reinterpreted. When trust is used as a device for establishing or maintaining social control, it becomes merely a means for furthering self-interest at the expense of others, expressed through an economy of domination. Distrust among people is "real." It does occur and has consequences in their self-understanding and relations with others. According to theological ethics, human dependency cannot be eliminated, but it can be taken up into a certain relationship that offers the hope for a restoration of a sense of trust in the meaning of everyday life. Humanistic psychology offers hope for a restoration of meaning and trust in life. However, from a theistic horizon of meaning, the estranged relationship caused by distrust is an illusion grounded in an alienated relationship with God.

A sociological and phenomenological review of the medical profession's fiduciary claims uncovers the roots of the use of trust-as-control by the profession in its attempt to retain power in order to fulfill its fiducial commitment. However, the use of distrust "in the name of trust" reveals a dynamic in which alternative interpretations and possibilities tend to be ignored or repressed. This limitation may provide momentary control, but it threatens to restrict the human experience of and participation in a fuller understanding of reality. As a result, there is the increasing temptation to encourage trust-as-control. In its commitment to understanding the faith-response to God, theology seeks to elucidate human understanding of God, the relation of God to the world, and consequently, the relations of individuals within the world. Thus, a theological argument can be developed showing the inadequacy of distrust as a basis for relationship since it leads to a perverse domination of the other (often under the guise of benevolence).

A theological interpretation of the "shock" that occurs between the finite provinces in the medical encounter serves to disclose the "hidden, yet revealed" meaning of new possibilities. For many in the postmodern world, a climate of uneasiness exists caused by a taken-for-granted investment in a particular understanding of God as transcendent, as "other." This perspective reflects the human tendency to interpret experience in terms of his or her own self-actualization and to forget the contingent, historically relative nature of meaning structures. This "image" or understanding of God provides a sense of reality which defines its limits and possibilities in particular ways. However, a re-examination of how we interpret meaning in our lives, and the implications of this process for social relationship, follows from a reinterpretation of the meaning of the divine-human relationship.

In a theistic perspective, God functions as an ultimate point of reference or orientation for all life, action, and reflection. As such, God cannot be conceived as one more item in ordinary experience or knowledge. God must be thought of as not restricted or limited by others. The concept of God functions as a limiting idea, in terms of which all else is to be understood, and which so transcends everything that it cannot be understood as gaining its meaning by reference to it ([151], pp. 12–13).[35] God is the idea of absolute aseity.

Charles Winquist and Mark Taylor, among others, have been critical of the metaphysical presuppositions that underlie this ontotheological conceptualization of God [267, 288]. Taylor argues, for instance, that a post-modern "death of God" has disrupted an assumed relation between essence and existence in which existence is believed to mirror the eternal ([267], p. 175; also see [58], pp. 667–668). This disruption breaks the fixed relationship of reality and appearance, and leaves humanity in the midst of pluralism and relativism. However, it opens the way for a theological hermeneutics which penetrates the assumption that a single, universal meaning is discoverable in theological discourse.

This disruption, this loss of a commonly accepted reference point, exacerbates the human feeling of uncertainty and dependency. Specifically in the condition of illness, it unleashes a struggle for control between the physician and the patient in which trust becomes a bargaining chip, jealously bartered between self-interested parties. The physician's desire to obscure the interruption of illness, to impose and maintain structures of order and meaning, is thwarted by the continual presence of "otherness." A theological contribution to a discussion of medical ethics begins with the recognition of the "boundary" nature of medical practice. The physician constantly must struggle to constitute meaning in a changing and fluid situation caused by the interruption of everyday reality which illness represents. Even if God's aseity is bracketed, the significance of God as a limiting idea remains a potent issue in reviewing the constructed or created meaning of trust.

In summary, some contemporary theological inquiries into the physician-patient relationship have been content to refute the medical profession's claim to special moral expertise and special prerogatives based upon their Hippocratic duty to benefit the patient. They seek to orient morality within the limits of the faith community [108]. However, whether one appeals to intuition, the consequences of actions, the idea of an impartial observer, or revelation and the nature of reality as God-centered, there appears to be no way to discover a universal, concrete character of the good life applicable over time and across communities. Nor is there a commonly acceptable method for delineating the goods, harms, duties, and obligations that occur in medical situations. Since one cannot be a fully detached, neutral observer, one cannot establish an universally accepted method to decide when one moral principle absolutely overrides another, or when those who deviate from one position are also the morally wrong ([70], pp. 33, 37). Since one must be a member of a tradition, its influence must be taken seriously. Those virtues that foster dialogical conversation among traditions or provinces of meaning become necessary.

Given an erosion of trust, and the resulting changes in the fiduciary context of health care, can theological ethics be brought to bear on the construction of a medical ethics which begins with an appreciation of the positive meaning of trust, and which recognizes the value of trust in the profession's accent of meaning? Answers to this question will vary according to the types of methodological commitments one brings to the task of theological ethics. I approach this work with a commitment to an hermeneutical exploration of the phenomenological and sociological structure of the medical profession's fiduciary commitment. This exploration reveals an avenue for the development of a medical theological ethics which articulates the distinctive contribution theology can make to the specific problem of the erosion of trust between physicians and patients.

In a theological reading of the text of relationship, it may be that the way the physician "cares" is as important as the "cure" he or she seeks. The fiduciary dimension that connects caring and curing must enable these concepts to evolve in meaning. Otherwise, an effort to maintain and control the paradigmatic "reality" of the biomedical model will obscure the important role of the caring and "sustaining presence" that is the correlate of cure.[36] By questioning or re-viewing the limits of the profession's horizon of meaning, we come to a renewed appreciation of trust and caring, and, perhaps, then are better able to cure. This review of the profession's horizon of meaning begins with its dedication to knowledge.

NOTES

[1] The patient-physician relationship is only one of a set of morally significant relationships in the clinical context. In this discussion, I am interested in the effect of unsettled trust specifically on the clinical relationship between patient and physician.

[2] Trust is not a property of individuals but is achieved interpersonally within a particular social and cultural milieu. The idea of trust adheres to an individual no more than sickness does; both are media of interaction. When we trust someone or when someone is sick, our interactions with that person take on a certain tone. It is in this sense that trust or its lack functions as a medium of social exchange (see [79], pp. 205–206).

[3] Indeed, the recent increase in government involvement, legal and insurance restrictions, auditing, and monitoring in health care has occurred because of the growth of distrust between the public and the medical profession.

[4] For a detailed discussion of the Christian perspective on illness and the biblical metaphor of the "poor" applied to health care, see ([253], pp. 55–90).

[5] For a discussion of the inordinate emphasis on individualism and the corresponding absolutization of autonomy in medical ethics, see ([179], p. 1132).

[6] A wide list of virtues necessary to medicine has been proposed through the centuries. For several excellent discussions of the meaning of virtue in relation to medicine, see [251].

[7] Barber uses the term "distrust" as an expression of the functional equivalent of trust in issues of social control ([8], pp. 21–22; see [173]).

[8] "Ordinary economic controls" refers to an organized alternative to the physician and to restraints on traditional market forces. Faith healers and alternative medical practitioners do exist, but are carefully proscribed by governmental regulations (see [262], Bk. One, Ch. 3; Bk. Two, Chs. 1 and 3).

[9] Parsons predicted that in postmodern society professional independence and lay control would prove incompatible; see quote in ([77], p. 515).

[10] For physicians the primary value guiding decisions is solidarity with co-workers, through a system of reciprocity. "Obligations to patients become a matter not of the relationship of the individual physicians to individual patients but of the house staff to each other or the house staff as a whole to a patient population as a whole. In this sense, the commitment to individual patients does not compete with commitments to abstract, potential patients but with commitments to other physicians who are friends and colleagues as well as indispensable allies. . . ." ([21], p. 215).

[11] Following the Nuremberg era, in which the integrity of the medical profession was tainted by the conduct of Nazi physicians, and coupled with the advances of scientific medicine, various groups within health care have attempted to ensure the public's trust in their ethical commitments. See the American Medical Association, "Principles of Medical Ethics" (1980); the American Nurses' Association, "Code for Nurses" (1976); and the American Hospital Association, "Patients' Bill of Rights," in ([13], pp. 122–127).

[12] As a style, "[p]ostmodernism is defined in the continual deconstruction of its definitions" ([79], p. 215).

[13] In a postmodern atmosphere, a professional's knowledge and fiduciary commitment are called into question. Consequently, if ethics accountability has required physicians to place the rights of patients among the values of medicine, this inclusion "may represent the triumph of form over substance" ([296], p. 171).

[14] For an excellent discussion of *strong* and *moderate* postmodernism, see [109].

[15] Husserl described this method as a "splitting of the Ego [Ich]: the phenomenological Ego establishes itself as 'disinterested onlooker,' above the naively interested Ego" ([136], p. 35). This same reduction, when applied to the interpreter, gave rise to the notion of the transcendental ego. Husserl developed the complex theory of the transcendental ego which is simultaneously the subject of his consciousness and the object of his cognition [137]. For a critique of this "splitting of the ego," see ([245], pp. 51–84).

[16] "Eidetic analysis aims at seeing through the particulars (concrete or existential) to discover that which is essential (ideal or typical). The essential structure of the phenomenon under study, whether it be social organization, social relationship, or belief system, refers to those elements which make up the phenomenon and without which it either ceases to be what it is or changes considerably. In an eidetic analysis we analyze the phenomenon, that is, our consciousness of the object, in order to discover its constitutive elements; then conceptual-

izations can be developed that assist us in discerning and understanding the empirical representations of the ideal-type" ([227], p. 10).

[17] S. Kay Toombs argues that physicians understand illness differently than patients and therefore exhibit traits which they feel are consistent with their horizon of meaning. Without an understanding of the patient's experience, doctors will work on a different plane [274]. Mary C. Rawlinson offers a similar argument in [230].

[18] Alfred Schutz argues that a person structures experience and organizes his or her "world" in terms of sub-universes of reality or "finite provinces of meaning" ([243], pp. 226–259). A finite province of meaning is defined as a certain set of experiences, all of which show a specific cognitive style – its "accent of meaning" (p. 230). The "habit of mind" is an approach to the world, the backgrounds and horizons of meaning against which and through which the particular province takes shape; see ([156], pp. 6–7). These provinces represent different worlds consistent within themselves but distinct from each other. All of the provinces are contained within the social matrix that Schutz calls the "world of daily life" or the "everyday world of working" ([243], pp. 208, 222). The medical profession is one of these "sub-universes" within the world of daily life. Christianity and other religious traditions represent various other sub-universes within the world of daily life.

[19] Toulmin believes that the disharmony in public attitudes toward the medical profession joins with a confusion of aims within the profession itself. As a result there is a shared uncertainty about the roles doctors are to play in our lives, the knowledge and understanding demanded of medical practitioners, and hence the quality of concerns that a medical "agent" should display toward any patient ([277], p. 32).

[20] I characterize these situations as "boundary situations." By boundary situation, I mean the point at which the horizon of known experience encounters the unknown; these are moments of uncertainty and risk. For instance, in medicine, no matter how knowledgeable or well practiced the physician, no matter how many tests have been run, there is the moment when the body is "uncovered" and the incision (or decision) is made. This moment of "crossing" represents a boundary situation (see [24], pp. 31–53).

[21] Focusing on the boundaries or limits of provinces places us "betwixt and between" taken-for-granted meanings, and exposes the risk or uncertainty that constitutes these moments of interaction. These situations expose hidden, yet present, meaning structures (see [267], p. 5).

[22] Stephen Toulmin argues that moral reasoning can give answers to why an act is right or wrong. However, there remain the "limiting questions" which may be asked – e.g., why should I be moral at all? In Toulmin's view, religious assertions have to do with these limiting questions (which can be asked when all moral answers and explanations have been given). Religion speaks to the condition of those who in actual living are afflicted by the sense of uncertainty [276].

[23] Religious faith itself occasionally may take on an illusory nature, requiring a transformation to a more authentic dimension. Soren Kierkegaard discusses the "leap" of faith in ([157], p. 262; 7:253).

[24] My interest is not with the truth of God's existence or nature but with the meaning of religious claims about God. Christian religious language tries to convey the "something more" that characterizes the religion. However, because these words are spoken in a world of rationality and human experience, they often appear paradoxical and contradictory. Generally, arguments at this point among theologians and between theologians and their critics are attempts to resolve the apparent or real contradictions contained in religious language. However, paradoxes do show something – if only that certain concepts do not apply when referring to God ([257], p. 43). For instance, to categorize God in modes of human experience would reduce or falsify the ultimacy that is associated with God. An appeal to supernaturalism, that God is just beyond any human understanding and that there is no necessary connection between the transcendent and the mundane, would seem to address this concern. Any claim of an experience of God could be explained, for instance, as a type of psychological event. Yet in the Christian tradition, it is explicit that there is

relationship with God. This problem of paradox, with all the epistemological difficulties that it entails of establishing the nature of this relationship between the indescribably transcendent and the manifest, has occupied the theistic tradition since its inception.

[25] *The Oxford English Dictionary* states that when applied to physics, *fiducial* can be taken as a standard of reference, e.g., a fiducial point around which other factors take their meaning. In the classical theological understanding of God, God can be viewed as a fiducial being – the point around which all else takes its meaning.

[26] The idea of God may be a human construct, subject to reconstruction, but this statement is not intended to mean that God is merely a construct of the human mind.

[27] "Otherness has entered, and it is no longer outside us among the 'others.' The most radical otherness is within" ([280], p. 78).

[28] The desire to eliminate otherness or difference becomes a drive marked by total absorption and control. It is in the nature of theocracy, for instance, that it must be willing to persecute in order to maintain its vision of a divinely ordained civic order and its claim to a divinely ordained truth that undergirds that order. The human tendency towards ideological totalitarianism reflects the desire to master the complexities of existence by dominating, if not destroying, the other (see [97]).

[29] The expression "living human document" is taken from [152].

[30] For a general discussion of phenomenology and sociology, see [167, 199]. Also, for a discussion of Schutz's methodology, see [9, 103].

[31] Conceptualizing medicine as a practice helps to situate the required skills within the framework of tacit knowledge, reflective practice, and a community of practitioners for whom certain values serve as boundaries for their work (see [183]).

[32] This professional view of the physician's authority has a certain legitimacy that must be acknowledged, even as it is critiqued. Legitimacy rests on acting in the patient's best interests; at least, in doing no harm. Often this claim to authority is grounded in the superior knowledge and experience of the physician in matters of health and illness. The structures of meaning within which physicians practices "place" them in the position of acting sometimes as if they know "best."

[33] Theologically, Niebuhr argues that the desire to deny our finitude is always inadequate and results in a sinful response to God. We must accept a relativism of faith, our social existentialism, and our freedom in dependence.

[34] Engelhardt's libertarian morality is one reaction to this perception of otherness [70]. The interest in autonomy reflects this concern. Autonomy requires an exercise of self-respect, self-concern, and self-control. Ideally, autonomy leads to relationships of mutual respect, benefit, and trust. However, realistically, autonomy fosters an ethic of mutual non-harm; see Erich Loewy's critique of Engelhardt in ([165], pp. 42–43).

[35] I accept Kaufman's assertion that ideas of God constitute a conception of the overall context within which human life is lived and in accordance with which forms of human existence are ultimately grounded. The basic thesis of his argument, however, is that theology finally is nothing but an imaginative construction of the human mind. It this sense, Kaufman's theology is reductionistic.

[36] Earl E. Shelp uses the expression "sustaining presence" in [252]. Stanley Hauerwas writes of a "suffering presence" in [123].

CHAPTER 2

TRUST AND THE MEDICAL PROFESSION

In a way, we enter a doctor's office already willing to trust the physician. As part of our assumption that physicians are trustworthy, we expect the doctor to adhere to an ethic of responsibility for our well-being. Attention to objective ethical rules, however, does not seem adequate to the whole experience of the medical encounter. After all, we expect doctors to "do the right thing" not because they are obeying a rule, but because they have developed a certain way of being that includes a sensitivity for persons in need and the knowledge to help. Although someone may not seek medical help for many reasons, generally when one does there is not a great deal of reflection or calculation required because we "naturally" assume that physicians know what they are doing and will act in our interests.

To sharpen the issues and to show the background of the argument, it is necessary to establish the meaning of "trust"; to outline the phenomenological dimensions of people's trust in physicians; and to establish a sociological understanding of what is meant by "the medical profession." This review will define the way in which the terms are used. It also will establish that a theological analysis of trust-as-control and trust-as-faith in the medical context must take into consideration the medical profession's social and historical nature.

THE MEANING OF TRUST

The place to begin an assessment of the profession's fiduciary commitment is with the meaning of the word "trust." As with the terms "profession" and "knowledge," trust and its seemingly related concepts – faith, confidence, distrust, and suspicion – are not well defined in much of the recent literature on professional ethics. "Trust" is used to refer to different things, and different words are used to refer to it. There are different psychological, social, phenomenological, moral, and theological conceptions of trust.

One attribute or characteristic used to qualify the physician as a status professional is his or her fiduciary relationship with a client. As an adjective, *fiduciary* conveys the meaning of "the nature of a trust." According to the Oxford English Dictionary, a fiduciary relationship involves a relationship "holding, held, or founded in trust." As a noun, a *fiduciary* is a person in whom confidence, faith, and belief can be invested because that person exhibits the quality of reliability and trustworthiness. The person or thing in which trust can be reposed becomes a locus of belief, expectation, and hope

([99], p. 33). A more careful analysis of trust's synonyms (e.g., confidence, reliance, and faith) and antonyms (e.g., distrust, uncertainty, and suspicion) further elucidates the meaning of fiduciary, and points towards "trust" and "trustworthiness" as definitional concepts in the medical accent of meaning. A review of the ways trust applies to the medical profession illustrates a variety of meanings associated with the term. In this discussion, the intention is not to evaluate the moral standing of trust. Rather, it is to develop an interpretive understanding the effect of trust on social relationship and social action. Therefore, it will be important to distinguish between confidence, faith, and belief as expressions of trust.

Trust is the concept representing the condition of confidence and expectation on which human life depends. If all people were invariably honest, and if there were no contingency in the world, we would not be concerned with the meaning of confidence and trust. We cannot live without forming expectations with respect to contingent events and, therefore, tend to bracket more or less the possibility of disappointment. The alternative is to withdraw expectations and live in state of permanent anxiety and suspicion. Both confidence and trust refer to expectations which may lapse into disappointments. Niklas Luhman suggests reserving the term "confidence" for "trust" when referring to the ability of a person, thing, or a social institution to function as expected [169].[1] For instance, we are confident that the car will start when we place the key in the ignition. We show confidence in a doctor's technical competence and ability to diagnose and cure us. Confidence grows out of the satisfactory confirmation of our expectations.

Trust, unlike confidence, presupposes a situation of risk and uncertainty. One can avoid the risk, but only if willing to waive the associated advantages within the situation. The public is encouraged to have confidence in the medical profession's knowledge and ability; trust becomes a concern in the experience of uncertainty accompanying illness. After surviving by-pass surgery, a person may have confidence in his or her physician's skill. Before surgery, however, the person must trust the physician's recommendation that the procedure is indicated, and that he or she is able to perform it. A second opinion provides additional information on which a decision can be based, but agreeing to surgery represents a declaration of trust in the physician. We do not depend on confidence as we depend on trust, although trust often is portrayed as a matter of routine and normal behavior.

Trusting behavior consists in action that increases one's vulnerability to another whose behavior is not under one's control and takes place in a situation where the penalty suffered if the trust is abused would lead one to regret the action [166]. It is based on an individual's belief as to how another person will perform on some future occasion, as a function of that person's current and previous claims, either implicit or explicit, about how they will behave. Therefore, this belief (which creates and maintains certain expectations) arises both before the individual can monitor such action (or independent of his or her capacity to ever be able to monitor it) and in a context which affects his or her own action.[2] A patient is not going to cut himself or herself open to see if the doctor actually performed the operation. The psychological notion of trust is related to the limits of our capacity to achieve a full knowledge of others, their motives, and their responses to endogenous as well as exogenous changes in the relationship. In this last sense, trust is a tentative and fragile response to our ignorance, a way of coping with the limits of our foresight and the freedom of others.

There is another dimension to trust. Trust also is directed toward and is invested in something outside of ourselves that elicits our response. Related to the Latin *fiducia* (trust or confidence), akin to *fides* (faith, also confidence), this term suggests that a fiducial relationship involves responsibilities founded on "faith," rather than on intellectual assent or knowledge. Under the Roman law of *contractus fiducia*, the partners to the contract were bound by a sacred oath.[3] Hence, a fiducial relationship or contract was a firm one, not to be violated without dire consequences. The parties to this relationship, therefore, were trustful and trustworthy or reliable, made confident by the nature of their fiducial pledge. Such a social relationship was imbued with an enlivening "power" rooted in the parties' mutual trustworthiness, by their involvement in a relationship which is more than the sum of its parts.[4] Trust, then, paradoxically, becomes the strongest and the weakest foundation on which reliance can be based. It appears to engender an instinctive, less reasoned, and more faithful reliance on its object than does confidence, which often suggests somewhat less definite grounds of assurance and is more easily affected by changing conditions.[5]

This position can be developed into an understanding of trust linked to the theological virtue of faith. Theologically, faith can be construed as trust or loyalty, and it represents a necessary condition for authentic knowledge of God and of the human good. A dedicated response to God provides the human with a context for understanding himself or herself and the world. This context provides a benchmark for discussions of good and bad in human life ([178], pp. 5–6). However, Julian Hartt argues that there is a tendency in modern thought to widen the gap between faith as belief and faith as trust. Accordingly, he claims that the cognitive mode is separated too easily from the dispositional.[6] Such pressure, Hartt seems to suggest, leads to rebellion against the notion of God as Lord and to the tendency to seek a locus of loyalty in persons or social institutions [203]. Here, faith can be monolithic and destructive of goodness. Strictly speaking, faith as a theological virtue introduces a new dimension into the moral life with its natural virtues, and can call into question even manifestations of religious faith. From a theological position, faith allows one to respond to the contingencies and exigencies in life. It enables one to find a stability and unity in life, an "authentic selfhood" which is centered in God and in relationship with others ([172], pp. 75–80).

Secularists would argue that stability and unity in life can be achieved in self-trust or trust in social realities of this world, that there is no need to speak of God at all. Religious faith, however, calls one toward a transcendence of the given, which is typical of self-critical religion. In the Christian onto-theological tradition, God is the ground of being beyond the immediate "reality," in relation to which life, with its dimensions of being and non-being, is given a meaningful basis. Ontologically, faith in this "other reality" serves to call into question faith as taken-for-granted belief in "reality."

Confidence, trust, and faith all express belief. Belief connotes the freedom of a subject to make commitments in the absence of full knowledge, with the possibility of changing belief upon receipt of subsequent knowledge. Confidence is a strong conviction or belief based on substantial evidence or logical deduction. Trust is an expectation or belief based on inconclusive evidence, and is tolerant of, indeed requires, the element of uncertainty or risk. Religious faith affirms the whole self by promoting reliance in "something more," "in spite of," even because of, doubt and uncertainty.[7] Trust thus occupies a stance in the middle of a continuum, between confidence and faith.[8]

The continuum is not bounded by unquestioning faith, on one end, and skepticism, on the other end. If skepticism is introduced, there would be no need for a discussion of trust. What people ordinarily mean when they speak of trust between doctor and patient is confidence in the professional's ability and social function. Trust, however, implies a reliance beyond such confidence. The fiduciary dedication of the tradition points towards reliance, but this reliance is subverted in a relation of trust-as-control. Doctors have encouraged faith or belief in the power, authority, and promise of scientific medicine, but in an era of AIDS, abortion, and the persistence of chronic illnesses, this encouragement fails to satisfy. The continued presence of these concerns calls into question the profession's ability to eliminate or control human finitude and suffering.

Inasmuch as trust is a dimension of all social relationships, trust first involves attention to order and disorder in society.[9] Second, actors' expectations are the starting point for distinguishing between different kinds of trust. As Bernard Barber argues, trust is not a function of individual personality variables nor of abstract moral argument, but a phenomenon of social and cultural variables ([8], p. 5). We have trust in someone's character, but character is shaped by social institutions and cultural values. Therefore, broad social changes can affect the different meanings and interpretations of trust. As I will argue, both physicians and patients have expectations of each other due to their differing accents of reality within the natural attitude of the everyday world. Physicians interpret the meaning of trust in particular ways due to their professional identity and their interactions with others. As these interactions change, their understanding of trust shifts, affecting the limits of their work.

Trust as a Form of Control

There is one more dimension of trust that requires attention. Drawing upon notions of Luhman and Harold Garfinkel, Barber views trust generally as the background expectation that the natural and social order will persist ([8], p. 9).[10] The more specific meanings of trust that underlie Barber's institutional analyses are the expectations that people will be technically competent and that people will carry out their fiduciary responsibilities. Taken together, these three expectations function to provide social actors with a sense of the world's orderliness, its cognitive and moral structure. Moreover, they also function as a mechanism of social control in regulating the exercise of power and as a mechanism of integration by creating and sustaining social solidarity ([272], pp. 89–90).

Barber argues that the logic and limits of trust suggest that when the balance of power between parties becomes unacceptable, distrust becomes the functional equivalent of trust in fostering social controls over the perceived dominant party ([8], p. 9). As Luhman points out, distrust is not necessarily the result of a breakdown in the social system. It can be a functional alternative and complement to trust as a form of social control.[11] Trust is the confidence in and/or reliance upon another. It cannot exist where there is absolute control over the other, but many acts are a mixture of trust and control. Barber argues that "distrust is not just the opposite of trust; it is also a functional equivalent for trust. For this reason only, is a choice between trust and distrust possible (and necessary)" ([8], pp. 21–22). Both trust and distrust serve to reduce social complexity and anxiety, although strategies of distrust convey negative connotations, and both act to shape and control social relations.

There are several limitations to Barber's argument, especially when applied to the medical profession. First, Barber basically limits his argument to the public's use of distrust to control the power of the professions. It follows, however, that a profession's members also could exercise distrust to control relations with their clients. The profession's cognitive and moral structure, its sense of the world's orderliness, its own social solidarity, and the value of its work depend upon a trust or expectation regarding its work in the world. This expectation may function as a mechanism of self-control in regulating its own exercise of power, but we must stress that the profession's "expectations" also regulate its relations with the public. Doctors may distrust those people or conditions that do not conform to their expectations.

Second, distrust may represent a functional means for sustaining relationships, but it is also a condition or process that can dissolve relationships. Particularly in the case of trust *qua* fiduciary obligations, Barber does not discuss how the loss of trust can be restored for the one whose trust has been shattered, e.g., the person who finds that the surgeon has recommended bypass surgery for lucrative fees rather than for actual physiological needs. Is trust once broken forever lost? This question is important when we examine the theological dimension of trust as a value in the medical relationship.

Finally, Barber fails to appreciate fully the complex and pluralistic nature of postmodern culture. Awareness of this condition highlights the issue of trust and distrust. As Barber argues, distrust is a functional equivalent of trust in the regulation and control of diverse people and groups in our society. However, he fails to recognize the extent to which this diversity draws distrust towards an anomic, paranoiac dysfunctional form. In the area of health care, for instance, the lack of a moral consensus is clear in the inability to resolve moral dilemmas and restore some sense of solidarity to social relations. This loss of a common expectation of orderliness and structure threatens to move beyond distrust as the means for the restoration of social balance, to a stronger emphasis on distrust, suspicion, and control as forms of domination between people.

In a move to reassert its social control over the profession, society has resorted to distrusting the physician. In this climate, trust comes to be used as a form of social control as each party in the relationship seeks to maintain its authority through a process of self-interested negotiations. The climate of suspicion and the role of trust represent changes in the physician's and patient's expectations of each other. Patients still "trust" physicians. However, what trust means to each party has shifted. Trust is no longer conceived of as a free gift of concern passed between the two that operates to establish mutually satisfactory parameters of an asymmetrical relationship. We will see that while distrust or trust-as-control does provide a structure for the medical relationship, it also transforms the experience of physician (and patient) into one of alienation from one's self and others.

David Mechanic argues that an erosion of public trust in physicians is due to the profession's drift away from its commitment to trustworthy behavior and beneficence to a concentration on profit and prestige ([185], p. 181). Physicians must be more trustworthy to regain the trust of the public. However, distrust or the use of trust-as-control also is used to restore confidence between participants in a social relationship. A discussion of trust's importance in shaping the medical profession's horizons of meaning and the physician's habits of mind broadens recognition of the importance of fiduciary commitment for medical work.

TRUST IN THE MEDICAL PROFESSION

The questions that emerge when one considers the relation of trust and the medical profession dictate my approach to this subject. Since it is usually assumed that the profession is primarily a fiduciary one, why should there be need of comment at all? Have the two become so separated that their relation is no longer obvious? Have the terms "trust" and "physician" become so ambiguous that their meanings are no longer clear, and is the attempt to address their relation actually a commentary on our uncertainty as to their purpose and function? Is it necessary to re-interpret to physicians the meaning of the medical profession's fiduciary commitment?

One thing is clear: When people are asked to say which professions command the most respect, they usually place physicians at the top of the list. Adjectives used to describe physicians include honest, reliable, and trustworthy. Doctors are seen as competent professionals, people of good character, whose priority is the good of the patient. Never in modern history has the medical profession been stronger.[12] However, at the same time, the profession's image seems to be suffering. Irwin C. Lieb has commented that "many of us have lost confidence in the ways such supposedly well-trained professionals as physicians . . . , among others, serve the public" ([156], p. 4).

Polls taken in the last few years show that the public is changing its perception of the medical profession. A recently published study of opinion polls demonstrated a substantial decline in public confidence in the medical profession from seventy-three to thirty-three percent [19]. There is much public questioning of professional trustworthiness. James Childress mentions a Harris poll in 1973 which discovered that fifty-seven percent of people surveyed trusted the medical profession, while fifty-two percent had confidence in local trash collectors ([41], p. 39). While seventy-three percent of Americans surveyed in 1983 saw physicians as up to date on the latest knowledge and technology, sixty-four percent of respondents felt that people are losing faith in doctors. In 1991, the AMA found that sixty-nine percent of the public said that people were losing faith [162, 211]. "[Over] two-thirds of the American public now believe that people are beginning to lose faith in doctors" ([185], p. 181). A Gallup poll taken in 1989 concluded that sixty-seven percent of respondents believe doctors are too interested in money and fifty-seven percent agreed that physicians "don't care about people as much as they used to" [*The New York Times*, Feb. 18, 1990].

These figures require interpretation, for when people are questioned about their experiences with their personal physician, they indicate confidence in his or her technical competence while expressing concern with quantity of contact time and increasing medical costs. Admittedly, it is difficult to measure patient "trust" in physicians. Polls generally elicit responses to categorical questions concerning physician competence, humaneness or personal qualities, efficacy of care, and availability [126, 135]. A response to these categories can vary with the wording of the questions.

There is no doubt that physicians possess special privilege in the public's eyes. However, there is a growing public concern that physicians are trading on their fiduciary reputation. An increasing number of people are disenchanted, finding that physicians are more interested in money or prestige and less concerned with their patients.[13] In 1982, forty-two percent of the public queried expressed the opinion that physicians' fees are usually reasonable. This figure declined to twenty-seven percent by 1984. During the same period, there was an increase from sixty percent to sixty-seven percent of those people who thought that doctors were too interested in money. To a great extent, physicians are coming to be seen as highly successful businessmen who are functioning "with the business ethic rather than the professional ethic" ([170],

p. 2879).[14] Since one of the basic features believed to distinguish professionals from occupational workers is the separation of client interest from personal and financial interest, public concerns with physicians' trustworthiness indicate a sense of distrust and uneasiness with the sincerity of physicians' "professional" dedication.

More directly, fear of the commercialization of physicians' knowledge, expertise, and power is shown by the issue of doctor referrals to laboratories or clinics owned by the physician or in which he or she has a financial interest. Common services in which doctors are shareholders include radiology centers providing X-ray and other diagnostic imaging tests, diagnostic laboratories, physical rehabilitation centers, cardiac rehabilitation centers, medical equipment leasing and sales companies, radiation therapy centers, hospital services such as operating rooms and laboratories, and same-day surgery centers [283]. Such arrangements are criticized because they produce a terrible conflict of interest for doctors, giving them powerful incentives to bend their professional judgments (see [140]). Given the vulnerable position of patients vis-a-vis the physician's knowledge, a suspicion that doctor's referrals rebound to the physician's fiscal benefit fuels the public's concern with physicians' patient-centered interest.

Because physicians are the holders and wielders of powerful knowledge, the issue of their trustworthiness is a social problem. An erosion of trust affects physicians' perceptions of themselves and their work. The more the public expresses concern, anxiety, and resentment, the more doctors are concerned and uncertain. There have been several reactions to this public questioning of their professional commitments. Many physicians and others hold that the idea of medicine as a business is disturbing, if not immoral [1].[15] There have been appeals to physicians to consider customer satisfaction and public relations to restore the profession's fiducial image [161, 285]. Others address public concerns with health care costs by concentrating on physician accountability and cost-effectiveness [194].

Some people argue that these concerns represent an implicit threat to medicine's professional mandate or an attack on the position of doctors as free agents. Medical practitioners assert that they cannot function satisfactorily when public distrust results in litigation and excessive regulation ([8], p. 131). They also object to "outside interference" in what they perceive to be the special prerogatives of their work, claiming that their professional codes already establish sufficient moral and ethical guidelines for their practice ([295], pp. 1–28; [193]). For example, many physicians look upon philosophical and theological analysis of their work as meddlesome and potentially debilitating to their care for their patients ([104]; [142], pp. 43–44).[16] Self-control of the limits of their work is viewed by physicians as necessary to their professional practice [82, 285]. Consequently, even when stress is placed on the moral distinctiveness of the medical profession's ethics, for many doctors this position represents a call for the restoration of physicians to their central and elitist position of control in health care [256].

The meaning of "central position" is subject to discussion. Social and economic changes over the last fifty years have served to question the physician's autocratic image as "captain of the ship." Previously, doctors enjoyed some latitude in shaping the content of health care decisions by gathering, defining, and dispensing information. The infusion of public monies into health care, the development of biomedical technology and therapeutic techniques, and the media's increased attention to these matters have made medical issues into public issues. More attention is given to the interests of various parties concerned in medical matters and to their respective beliefs and ways of reaching decisions. As a result, the doctor's authority competes with the increasing emphasis on the patient's autonomy and the professionalization of other health care occupations. Since some form of power and control is necessary in social relationships to set goals, direct work, and establish order, the competition between physician authority and patient autonomy raises the issue of trust in health care ([8], pp. 19–24).

Sociologically, trust is necessary to social relationship since it is the basic ingredient without which the relationship would falter. However, trust is not an absolute feature of relationship – trust depends on a situation of risk in which the other person or thing is not under one's control. It is commonly believed that doctors know what is best for their patients, and that, therefore, patients should trust them to act in the person's best interest. In reality, the relations between physician and patient are a mixture of cooperation and self-interest, and trust must be created and maintained in an on-going relationship. Ideally, this relationship is one in which trust is given and received freely. However, when uncertainty and suspicion are present, relationship becomes marked by self-concern and self-protection.

This issue is not of only speculative or abstract interest. Practical problems arise in particular social and intellectual contexts within the world of working. Medical theory and practice are guided by multiple values and paradigms that sometimes conflict – both within medicine and with values of the multiple groups outside the boundaries of the professional view – and shape the expectations which physicians and patients have of each other.[17] Of immediate importance, the uneasiness caused by such conflicts disrupts the social interaction that forms the basis for the medical relationship.

As the public image of physicians has suffered in recent years, the medical profession's understandable reaction is to examine ways to restore its image and prestige. While the profession has begun to appreciate certain criticisms of the biomedical model of illness, has begun to appreciate the calls for patient autonomy and rights, and has revised its codes to reflect these concerns, the depths of the issue remain unexamined. These efforts serve to maintain the authority of the profession. The "shock" caused by public distrust is ameliorated by incorporating the conflicting concerns into the physician's accent of reality, thereby restoring his or her sense of control, power, and authority.

A SOCIOLOGICAL UNDERSTANDING OF THE MEDICAL PROFESSION

Few scholars agree on the characteristics that determine professional status. In a broad sense, a profession is an occupation for which one is paid. However, all occupations do not qualify as professions. In a narrower sense, a professional is one paid for skill and expertise in an area of specialized work.[18] In the last century a range of activities and occupations have claimed professional status. Plumbers, dry-cleaners, and baseball players are among the many groups which describe themselves as professionals. Yet while baseball players and dry cleaners are paid for their skill, few would grant them the same respect as physicians. If people regard a particular occupation as a status profession, scholars have assumed that there must be a fundamental quality of that group which sets it apart from occupational professions. Often it is argued that this quality is a dedication to service or a fiduciary commitment to the client.[19] In examining this issue, it is necessary to review attempts to define the meaning of "profession," particularly in relation to the medical profession.

Three theories specify the medical profession as a status profession: the traits theory, the function theory, and the institutional structure theory. The first theory takes as its premise the assumption that those who are members of the medical profession reflect a set of common attributes that distinguish them from other professionals or from non-professionals ([3]; [191], pp. 51–65; [86]). The trait theory claims that there are identifiable attributes that form the representative core of a professional occupation. Therefore, this approach conceptualizes a status profession as an inherently distinct occupation, distinguished by essential qualities that can be found in other status professions, although these qualities may vary in importance among them.

The trait approach is an inadequate theory because it implicitly accepts the premise that there are "true" professions that exhibit all of the essential elements.[20] The problem with this approach is that "essential" is defined by abstracting from the known characteristics of existing professions. Because there is no commonly accepted definition of an ideal profession, the theory does not explain why trust is necessary in medical work.

The second approach identifies particular patterns of social behavior which distinguish physicians from other members of society. Those who favor function theory argue that the physician is characterized by the role he or she plays in society, and that this role is consistent over time and in various situations.[21] According to this theory, occupations become status professions when they function as controllers of and gatekeepers to specialized knowledge and skills. Because of the gap in competence and knowledge between professional and client, to fulfill its role the profession must be dedicated to acting in a disinterested way for the benefit of its clients.[22] This commitment to knowledge and objectivity serves to elicit the trust of the public in the profession. It also serves to maintain social equilibrium by controlling access to and removal from the "sick role."[23]

As a member of the medical community, a functionalist would argue, physicians possess the theoretical and technical knowledge necessary to fulfill their functional role. However, their knowledge is not unlimited or absolute. To pursue its social function, the profession must dedicate itself to the continual pursuit of knowledge and expertise. Thus, another functional component of the profession becomes the application of scientific knowledge to clinical practice ([216], pp. 431–432).[24] Also, to fulfill their role physicians must cultivate objectivity as a guard against any emotional involvement that might cloud their professional judgment.[25]

Functional theory offers some advantages over the trait theory of professions. Traits ascribed to an "ideal" profession become secondary to the social function played by the professional. According to the functionalist perspective, a dedication to beneficence is attributed to physicians because it enables the doctor to gain the patient's trust, thereby better enabling the doctor to restore the social equilibrium by removing the patient from the sick role. Therefore, the ranking of traits, or the inclusion of some and the exclusion of others, may vary over time, depending on how efficacious they are in supporting or structuring this function.

Just as trait theory attempts to characterize normative qualities of professions, however, functionalist thinking treats the medical profession as a generic concept rather than as a changing historic concept with particular roots in industrial nations influenced by Anglo-American institutions.[26] The assumption that all status professions at any time and in any culture have similar roles suggests that a common function, with common norms, can be abstracted or distilled ([87], p. 31). However, if no normative function can be elucidated, functional theory is explanatory, not descriptive, of the medical profession. Such a view undermines the assumption that physicians are a homogeneous group sharing a unique, univocal character.

Recent work in the sociology of professions argues that professions are social organizations whose structure and relations with other social groups vary over time. In this view, professions are distinguished by certain social dynamics of interaction that give them access to power, an institutional structure that evolves to sustain that position, and activities by which they exercise that power ([87], p. 16). Traits, then, are reflections of ideals that underscore the profession's power. Their variability is not as important as the way in which they are used to realize professional power and authority. Power enables and authorizes physicians to play a social role around which certain traits accrue. These traits enable physicians to control their work in society. The way in which a social institution controls its work, therefore, influences conceptions of its function. The third theory states that the medical profession is a social institution which is able to control its relations with others in the society (both those within and outside of health care).[27]

If we assume that medicine is a social institution essentially similar to other social institutions, it develops, changes, and adapts historically, in response to social, economic, and political changes [75, 144, 291]. According to this focus, medicine is an occupation that has most effectively gained and works to maintain a monopoly in its market. Jeffrey Berlant argues that medicine developed as a commercial organization to achieve control of its "market situation" ([18], pp. 51–58). To gain power and control of its art, the medical profession developed an organized structure, developed an institutional ideology, limited access to its ranks, and claimed "status" as a profession acting in the best interests of its clients [262]. This process results in professional solidarity, protects the profession from competition, and fosters a high degree of autonomy against lay control.

The drive for institutionalization creates a condition of uncertainty and distance between the expert and the layperson that reinforces the demand for professional autonomy ([145], pp. 42–44). First, an expert's high degree of competence and knowledge means that lay people cannot judge the quality of professional performance. Accordingly, a patient must trust the doctor's knowledge and skill, and subordinate his or her own autonomy to the doctor's authority. Second, medicine claims that its professional organization and structure efficiently work for the client's benefit. As the sole legitimate arbiter of proper performance, the profession argues that this autonomy enables it to serve its clients most effectively, efficiently, and safely. Thus, self-regulation through codes of ethics and informal mechanisms of control through education, training, referrals, and peer review play a major role in the medical profession's claim to autonomy [54].[28] The medical profession sees itself, and works to be perceived by the public, as a service occupation dedicated to the society's central values. Thus, the profession's "profession" of service is tied to power and control over members' work, while supporting physicians' autonomy and authority in society.

Trait and function theories represent a static view of the medical profession, the first by seeking qualities possessed by the "ideal" physician, the second in describing a gatekeeper or mediator role played by physicians. The third alternative description of the profession as an institution of power dismisses the discoverability of essentials, and suggests that a profession creates its identity and purpose in an ongoing struggle of social dimensions. However, such a description still suffers from the deficiency of the earlier sociological theories. Although it emphasizes the process by which occupations become institutionalized or professionalized, it still presupposes the idea of profession. Without some prior definition of profession, the concept of professionalization is meaningless. It is impossible to escape this definitional dilemma if we assume that "profession" is a generic concept discoverable as a universal, univocal idea ([87], pp. 29–36).

To summarize, it is difficult to find agreement on a definition of the word "profession." The word is evaluative as well as descriptive. With this difficulty in mind, the medical profession can be defined as an institutionalized occupational-collectivity constituted by a formal body of theory, knowledge, and technical expertise, able to exercise power in society through control of its work. It gains power and prestige through control of its training and socialization, the development of expert technique, and its claims to maximum efficiency in delivering the services it represents. The profession's social and organizational structure "constitutes" the sense of reality that individual physicians share, and with which they approach their work. It also has developed and encourages a fiduciary commitment towards its clients that enables its members to perform their work (and earn a living). Phenomenologically, much of the structure and process by which the profession has gained and preserves its power is taken for granted and remains unexamined. This "control" influences the way in which a physician conceptualizes his or her fiduciary commitment to medical "work."

Knowledge and Trust as Definitional Characteristics

Two of the necessary components delineating the definitional boundaries of the medical profession, and the identity of the physician who is a member of the profession, are expert medical knowledge and a fiduciary commitment to the welfare of the patient ([87], pp. 21–24; [148], p. 1306). The two foci balance each other in a reciprocally related fashion. Society allows medicine to develop its knowledge and skills, and grants the doctor great autonomy and social status, in return for the assurance that the doctor can be entrusted to use his or her knowledge only for the benefit of the patient. The long, arduous course of mastering medical knowledge and technique imparts to the physician a confidence in his or her ability, skill, and expertise. This expertise also serves to separate the physician from others and, in part, qualifies him or her as a professional. Along with expertise, trustworthiness becomes a key feature defining the meaning that the medical profession represents to the public. All doctors may not act disinterestedly (indeed, acting in a fiduciary manner can be in their self-interest), and an altruistic or benevolent concern may be a common moral duty to all people, not just professionals. However, in the relations between doctor and patient, moral concerns are emphasized since there is a great disparity in expertise and knowledge between doctor and patient. The physician comes to trust in the value of his or her work and in his or her own trustworthiness.

The two components of knowledge and trust also establish the nature of the physician-patient relationship. Physicians' medical expertise and knowledge, coupled with the patient's condition of illness, creates a disparate balance of power between them. Physicians and patients live in a common social world in which their assumptions about the asymmetry of their relationship are largely shared and have developed over time. By virtue of the authority vested in their professional role, physicians control the patient's access to and understanding of medical information and their expertise. In this process, they function as "gatekeepers," providing options to some and denying them to others. This gatekeeping power requires that the public trust the physician to act benevolently in the patient's interests.

Ordinarily the public accepts this asymmetrical relation due to a need for help and a confidence in the professional's dedication to a fiducial interest in the client's well-being.[29] However, the profession's fiducial identity is of reciprocal importance to the profession. The physician's authority and power is closely tied to the assumption of trustworthiness by the public. For the physician, the commitment to trustworthiness shapes his or her perception of what it means to be a physician and becomes a virtue defining the value of his work.

It is an inescapable feature of the medical situation that other people's ill-health, misfortune, and disadvantage are sources of power, status, and income for those persons in society who offer their services as physicians [35]. After all, the medical profession is an occupation, and the physician is a member of the economic division of labor. Therefore, the doctor has a self-interest in medical work. Because the patient's vulnerability makes him or her susceptible to abuse by the physician's self-interest, the relationship is one with moral connotations. Again, the way in which the physician understands and incorporates trust and trustworthiness shapes his or her self-conception. It has important repercussions for the nature of medical work. The physician is not encouraged to see this work as a "business" or to use the patient for his or her own gain. Rather, the profession and the public take for granted that the physician is a moral person dedicated to the patient's interest. Thus, the patient and the physician share a reciprocal (albeit asymmetrical) social world in which the physician's authority ordinarily is assumed or accepted by both parties.

Most attention to the moral and ethical dimensions of medicine continues to focus on the needs and position of the patient. I do not deny that physicians assume a dominant attitude in regards to their patients, or that patients tend to accept a reduced or diminished status in an asymmetrical relationship. A corrective to this unequal relationship has been necessary. However, attempts to articulate and enhance the patient's position work to limit the physician's dominance.

By recognizing the way the profession is constituted by a series of social relationships, the foundation is laid for an understanding of the importance of trust in the medical profession's definitional boundaries. I do not mean to argue that medicine's interest in knowledge and service are reducible to social dynamics and power-relations. Since the physician and patient are involved in an asymmetrical relationship, issues of power, authority, and control raise moral concerns. Each participant interprets and understands "physician" and medical "practice" from within his or her own horizon of meaning.

In chapters three and four, I will argue that physicians experience their profession as a horizon of meaning, within which knowledge and fiducial commitment to service are important values to physicians. At the same time, there is no final, essential definition of "physician" that can aid in discovering a basis for professional morality compelling to all rational agents. It is important to recognize the institutional and historical nature of the profession as an organization seeking to control its work. The physician, as a member of this organization, acquires the profession's values and meaning structure, which shape the doctor's identity and practice. The physician is involved in interactions that require attention to the issues of power and authority in social relations. The historical dimension of the profession is important as well. Institutions are not monolithic or universal; their structure evolves over time. This evolution is created by the shifting power structures within the evolving institution and its shifting relations with others in its historical era. Values play a part in shaping this dynamic. They also can be affected by it. Expert knowledge and a professed dedication to trustworthiness function to help doctors control the limits of their work. If used in this way, however, these values become limiting to a relationship of mutual dependency.

NOTES

[1] Although he does not use the term, Luhman seems to associate confidence with "habit." As long as a person, institution, or thing functions as expected, we tend, without direct or deliberative reflection, to take it for granted; we interact in a habitual fashion.

[2] Another case where trust comes into play is when others know something about themselves or the world that I do not, and when what I ought to do depends on the extent of my ignorance of these matters. An agreement between myself and these others may call upon them to disclose their information, but can I trust them to be truthful (see [55], pp. 51–52]). A patient must trust the doctor, i.e., rely on the doctor's actions under conditions of identifiable and unidentifiable risks. Also, the notion of trust has something to do with relationships of mutual dependency. Unfortunately, this sense of mutual dependency often is transformed into patient dependency upon the physician (see [119], pp. 184–202).

[3] More specifically, the fiducia was a contract of sale to a person by mancipation, coupled with a sacred agreement or oath that the purchaser should sell the property back upon the fulfillment of certain conditions. The oath became the basis for trust on which the exchange relationship was founded [286].

[4] It is this sense of trust or faithfulness that is represented by covenant relationship as opposed to contract in medical relationships. Contracts, today, generally are made because of a lack of or insufficient basis for trust. In a contract arrangement, the "parts" of the relationship are carefully described and proscribed. For a summary article on the covenant-contract discussion, see [46].

[5] Trust provides an element of certitude that extends beyond the vicissitudes of momentary experience; C.f., Heb. 11:1; Ps. 141:8.

[6] "[Faith] as trust, its ties with certifiable knowledge of God and of the human good loosened, is put under heavy pressure to posit in and for itself value absolutes – ideals and/or beings worthy of unconditional loyalty" ([117], pp. 222–224).

[7] Faith is an emotionally charged condition which in the extreme becomes an unquestioning acceptance that excludes doubt. However, we cannot do without some perspective, and therefore faith rests upon interpretation and relationship ([115], p. 63; [202], p. 118).

[8] Such a definition tells us that trust is better seen as a threshold point, located on a probabilistic distribution of more general expectations, which can take a number of values suspended between complete distrust and complete trust, centered around a midpoint of uncertainty. Accordingly, either blind trust or distrust represents the predisposition to assign the extreme values of probability and maintain them unconditionally over and above the evidence ([55], pp. 51–52).

[9] Trust reduces complexity in social systems [168]. Trust is a "social good. . . . When it is destroyed, societies falter and collapse" ([20], p. 26).

[10] This sociological argument rests on the ontological presupposition that there is order upon which the natural and social world depends, a presupposition that is necessary to make

talk of trust meaningful. For a discussion of the ontological presuppositions or foundations on which society depends, see [287].

11 Luhman echoes Parsons's systematic analysis of trust and distrust in social relationships; see [169]. The dialectical relationship between trust and distrust operates like the "carrot" and the "stick." The use of one without the other will not be as effective as the use of both.

12 George Lundberg claims that modern doctors are better trained and more competent to deal with practically every kind of treatment problem and prevention strategy – and wonders why the profession is undergoing reappraisal [170]. David Mechanic sees the growth of public disillusionment accompanying magnificent medical advances [185].

13 A 1984 survey by the AMA indicated a growing public perception that physicians were becoming too interested in money, spending less time with their patients, and expressing little interest in them (see [89, 263]). The medical profession is concerned with these perceptions. In the last ten years applications to medical schools have declined. One reason cited by Marvin Dunn is that college students ". . . began to see medicine as more of a business than a profession." For this reason, he concludes, physicians must work to preserve the values that define medicine as a true profession [62].

14 Businessmen might express dismay at Lundberg's disassociation of business and professional ethics. Also, for a comment on the business aspects of medicine, see [25].

15 For an alternative position, see [71].

16 Some physicians and others argue that bioethics actually interferes with "good" medical care [Bennett, 1980]; [Greenberg, 1974].

17 After all, the manner in which the doctor conceives, thinks, and views "his own knowledge and understanding directly shapes the manner in which he perceives, thinks about, and treats the patients who are the objects of that knowledge" ([277], p. 33). The same is true for patients. If physician and patient have contrary ideas about the character and claims of medical knowledge, they are likely to end up with conflicting notions about the nature, terms, and mutual obligations of the professional relationship.

18 Freidson describes medicine as one of the "learned" professions that, traditionally, are the occupations of the educated and high born ([87], pp. 20–35). In modern usage the word "profession" is often defined in terms of technical expertise sold for a fee, but the more traditional meaning describes a group that professes a vow to an ideal of service [64]. The word profession comes from the Latin *profiteri* which means "to declare aloud," and a professional may be defined as one who "professes, who takes an oath, a scared vow" [52]. For further discussion of the "professional" nature of medicine, see [31].

19 If we distinguish between status professions and occupational professions, occupational professions are those groups whose members use their knowledge and skill primarily to make a living ([67], pp. 14, 32). A status profession is one which is given qualitatively higher status because it is assumed that financial reward is a secondary concern to its members ([7], p. 671).

20 Trait theory assumes that a profession is an organized community, based on shared identity, values, and role definitions, whose authority is accepted by the public [100]. For a critique of trait theory, see [30].

21 William May, Talcott Parsons, and Bernard Barber are scholars who use a functional or role method in describing the profession's defining characteristics (see [176]; [215], pp. 34–49; and [7]).

22 In reviewing Talcott Parsons's work on the physician-patient relation, Renée Fox discusses the importance of the "competence gap" that exists between physician and patient which binds the two parties together in a semicollegial relationship ([77], p. 500).

23 Parsons sees illness not merely as a biophysical condition. It is an integral part of a social process that is governed by the institutionalized roles of the medical profession, which has as its particular function the "management" of disease and illness. His analysis of medical practice also recognizes the physician's semicollegial relationship with the patient whose "sick role" is complementary to that the physician ([220]; [216], pp. 428–479).

[24] Parsons stresses the innate relationship of medicine with science. The use of science to bolster knowledge, and in turn better enable doctors to perform their function, increases the impact of the argument that scientific medicine and technology improve the doctor's ability to act for the patient's good. Parsons sees the medical profession as the embodiment of the "primacy of cognitive rationality" ([217], pp. 536–547).

[25] For a description of the particular institutional conditions which lead to the development of objectivity as a characteristic attitude, see [14, 187]. In order to function effectively, the doctor must establish his or her authority through technical expertise and affective neutrality. The primacy of cognitive rationality in medicine allows for functional specificity and affective neutrality. Therefore, an emphasis on technical roles in therapeutic relations serves to exclude considerations that would undermine the efficacy of physician's function and impede performance. The alliance of science with medicine fosters this combination of technical skill and affective neutrality, thus actually reinforcing trust in the therapeutic relationship ([145], pp. 35–36).

[26] Johnson claims that the functionalist emphasizes the gatekeeper role as ahistorical and homogeneous among (status) professionals ([145], pp. 36–37). Also, Berlant critiques Parsons's functionalist theory of the medical profession by noting that Parsons claims to present a descriptive argument. Minimally, however, it contains an implicit explanatory theory for the institutionalization of the profession's normative structure ([17], p. 12).

[27] Johnson and Freidson argue that the functionalist perspective is too unilinear in concept. Instead they favor a social analysis of professions as social institutions. Professions are those collegial occupational groups that exercise control over their work (see [83, 85, 145].

[28] Daniels notes that, for the profession, the autonomy of the physician is not desired out of self-interest, but is a requirement for offering the best possible service in the public interest ([54], p. 39). Such an argument, echoing Parsons, is used as an apologia for public trust. However, "it may be that the characteristic claim of service to mankind which professions make is as much an unquestioned assertion that everything and anything a professional does is by definition service to humanity as it is an assertion that professionals are obliged to determine what it is that does serve humanity and how they might better strive to do so" ([84], p. 13). Therefore, codes function as a device to retain or advance monopolistic control of work, as a means for receiving government sanction of an exclusionary shelter in the marketplace. Knowledge, skill, and service orientations are no longer regarded as objective characteristics of the institution of medicine. These characteristics become ideological devices to establish autonomy, as the means to gain or preserve social status and privilege (see [85]).

[29] Patients have limited access to the medical knowledge and technical skill of physicians, and limited ability to question the need for medical procedures. They are dependent on their physicians' judgments. Thus, it is important for patients to be able to trust their physician ([73], p. 4). It should be noted that physicians are caught in a double bind. They are given the social function and responsibility to make medical decisions. However, they are dependent on their patients in many ways: for truthful medical histories, for compliance with medical directives, and for the opportunity to practice the art of medicine. Therefore, it is important for a physician to be able to trust his or her patient.

CHAPTER 3

EXPERT KNOWLEDGE AND THE MEDICAL PROVINCE OF MEANING

The medical profession qualifies as a status occupation because it is a social institution wielding great power over its work. Certain traits associated with physicians express, as well as perpetuate, the profession's power and function in society. These traits may change from time to time, and the function played in society by doctors may be reinterpreted. However, for status professions such as medicine, the possession and control of expert knowledge is a consistent characteristic. Expert knowledge is deployed in different institutional forms and used as the basis of individual and collective power and privilege. The development of additional knowledge is a natural consequence and leads to further efforts to control the expertise on which the power of the profession rests. To a great extent, the parameters of medical "work" are created by the possession, development, transmission, and control of theoretical and technical knowledge. The importance of knowledge to the physician's work plays a part in delineating the values attached to that work by physicians.

Of course, many occupations are groups of people sharing certain relatively arcane knowledge and skills. Therefore, there are two foci consistently stressed as characteristic of status professions: the possession of theoretical and technical knowledge, and the expectation of a fiduciary relationship with a client. The medical profession has a long tradition of espousing service to clients as part of its self-identity and its identity in the eyes of the public.[1] The reason for the fiduciary focus is the knowledge gap between expert and lay person, and the possibility of harms to the client if the physician uses this power for personal gain or aggrandizement.

In this chapter I will examine the component of knowledge from a phenomenological point of view, building upon the sociological work of chapter two. I will explore the *like-mindedness* that serves to distinguish physicians' work, developing a model or typification of the medical profession as a "finite province of meaning" grounded in the biomedical model of illness.[2] This discussion reflects an epistemological issue: What are the structures and limits of medical knowledge? Second, it raises a question from the sociology of knowledge: How does a commitment to developing, controlling, and applying medical knowledge affect the physician's work?

It is the second question that forms the primary basis for this chapter. A phenomenological analysis requires that focus be placed not on the specific organizational structure of the medical profession, but on the way in which its members interpret their own organizational world (a special sphere of the individual's *Lebenswelt* or everyday life-world).[3] The phenomenological concept of the life-world refers to the fact that in any real-life experience, something is pregiven, taken for granted, and passively received by a person's consciousness. This taken-for-granted world includes our cultural world, and whatever prejudices and interpretations may derive from it. We also approach the moral dimension of this world with a prior set of understandings that require us to continuously reconstruct the moral meaning in our lives. Organization of experience into typical forms from which we derive our moral sense of the world is an important part of this ongoing reconstruction. The "typical" is a deeply rooted feature of the organization of experience into knowledge. The medical profession achieves, in part, its power and authority from control of the knowledge, actions, and values necessary for its work. Therefore, the concept of "typification" is useful in depicting the way in which physicians "know" or experience and value their world, their work, their relations with patients, and the way in which they see themselves in relation to these concerns ([243], p. 7).

If commonsense knowledge of the everyday world is organized and distributed according to structures of meaning, which are pregiven and shaped by typifications, then the question of the structure and limits of medical knowledge implies several other questions: What is the role of the profession's social structure in the development, control, and transmission of medical knowledge and expertise? What social relations are presupposed by the forms and structures of medical knowledge? Finally, what part does expert knowledge play in shaping the "horizon of meaning" or standpoint of the physician *qua* professional?

Also, if physicians' understanding of knowledge contributes to the way in which they find their work meaningful [277], how does this knowledge shape, and how is it shaped by, the limits of the profession's autonomy, power, authority, and social relations with the everyday world? The basic epistemological model by which physicians make sense of their work is the biomedical one. If there is a correlation between the social structure of the profession and the system of knowledge it endorses and reflects, the emphasis on scientifically grounded knowledge shapes the social relations of the profession with its clients.

The biomedical model is one particular typification of medical knowledge and derives from an investment in a scientific approach to medicine. The development of the biomedical model of illness and the concomitant development of medical technology reinforce the adoption of a scientific ideology by the medical profession. Given that typification is a natural component of the life-world, the biomedical model matches closely the structure of typicality that is found in medicine. In fact, many physicians assume it to be the natural model outlining the limits of medical work. Learned in medical school and reinforced through the institutional organization of medical care, its use shapes the physician's *accent of reality*, thereby affecting the nature of the fiduciary component of medicine.[4]

KNOWLEDGE AS A DEFINITIONAL COMPONENT OF THE MEDICAL PROFESSION

The modern health care system is a complex network of practices growing out of different historical traditions, embodying different values and different methods. For the people involved in it, the system weaves together ways of knowing, acting, and valuing that constitute health care. As members of this amalgamation called a system, physicians operate within this interrelationship and contribute their practice to it. We will focus on the important role played by "knowledge" on physicians' actions and values.

A number of modes of knowledge operate in medicine. As Toulmin writes, "In medicine, more than any other discipline, our task is . . . not to define 'medical knowledge' restrictively but to recognize the plurality of different types of medical knowledge" ([277], p. 41). Perceptual knowledge of the external world, knowledge of the "other," common sense, technical, political, philosophical, and scientific knowledge are all types of knowledge with social or intersubjective importance ([111], pp. 23–36).[5] In medicine there is a correlation among these various types. All play a role in some combination of theoretical (both philosophical and scientific) and practical (both empirical and clinical) knowledge. All serve to delineate the boundaries of medicine, and therefore, become aspects of medical knowledge (directly or indirectly).

Generally speaking, in medical knowledge, this combination is focused on a particular patient's clinical treatment. The doctor's understanding of medicine relies on general physiological or psychological principles that have medical significance only insofar as they can be related to a personal understanding of the particularities of clinical practice with actual patients. Thus, the distinction between theoretical or formal knowledge and practical knowledge is reproduced in the medical field as the distinction between biomedical science and general clinical medicine ([277], p. 40). The practice of medicine relies on a coalition of different theoreticians, researchers, specialists, and general practitioners, for whom knowledge is a different commodity and the patient a different "object" of knowledge.

For our purposes, "medical knowledge" refers primarily to the formal knowledge of general biomedical science, to technical information, and to factual information grounded in the empirical method on which the practice of modern medicine is based. In addition, it incorporates the knowledge gained from the physician's experience in everyday work. Medical knowledge is in large measure a constructed competence that results from formal reflection and practical experience. It includes generalized intellectual skills, the various areas of knowledge outside of any one specialty, skill in handling interpersonal relations, and "common sense." Thus, such knowledge derives from many sources and can cover a range of applications.

Medical knowledge is a complex, evolving, dynamic concept, both theoretical in general and specific to the circumstances and demands of each clinical encounter. Medical knowledge might be envisioned as stretching over a continuum from a research orientation with no direct patient contact (although experimental involvement is possible) to the immediate knowledge of the condition of a particular patient. Because the profession is oriented around the relief of suffering, all medical knowledge rests finally on providing service to the client.

It is with this last dimension that the issue of trust becomes focused. In any age a patient must trust that his or her physician possesses the theoretical and technical knowledge that the patient does not possess or understand. The context of this confidence is most immediately pertinent in the clinical encounter between patient and physician when the professional's knowledge is put into practice. The nature of medical knowledge and the goals to which it is applied both determine and reflect the spirit in which patients are concurrently treated. Thus, biochemistry may be important to the clinician, but the knowledge gained in the laboratory is several times removed from the examining room and the patient. It becomes important to the degree that it has an immediate relevance to the concerns and questions of the patient. The physician is responsible for knowing the larger parameters of his or her field and the way to apply this knowledge to the needs of any one patient. The emphasis placed by the profession on a scientific basis for medical knowledge largely determines and reflects the spirit in which patients are "treated." Even the commonsense knowledge of physicians, as part of the stock of knowledge that doctors bring to their work, is shaped by and placed within the structure of formal and empirical knowledge.

As a social institution, the medical profession lays claim to a reservoir of knowledge gained from its tradition, a commitment to biomedical science, and a long history of practical experience. This knowledge is organized, evaluated, and transmitted by means of a process characteristic of medicine. Medicine "classifies" and "evaluates" knowledge according to the definitional boundaries of the field. Knowledge becomes a factor in the profession's social, economic, and political power to define and control its work.

Medical knowledge is dedicated to the values of restoring and maintaining health and relieving suffering.[6] However, the expert knowledge that hopefully enhances these goals also serves to separate the physician from those people who come to him or her for help. The public is not expert in medical matters. Physicians are expert in comparison to their patients, who know very little about medical work, though they may very well know a lot about the experience of illness ([212], p. 15). Through education and socialization into their profession, doctors have modified their understanding of the everyday world – and the experience of illness – to develop a particular or *typical* knowledge of the life-world. Such typical knowledge implies a consensus, shaped by an accepted base of knowledge shared with colleagues, that determines the understanding the physician brings to concrete situations. The acceptance, the utilization, and the modification of this knowledge invests the physician with the power of his or her expertise. This expertise provides an *accent* to the reality of a physician's work in the world by organizing experience according to coherent patterns of meaning.

As we have seen, if the natural attitude of the world of everyday life is bracketed by employing the phenomenological epoché, multiple or finite provinces of meaning are revealed.[7] A finite province is defined as a certain set of experiences, all of which show a specific cognitive style or accent of reality ([243], p. 230). Examples of these provinces are the world of physical things, scientific theory, religion, madness, music, and medicine. Each province represents a different social construct consistent within itself but distinct from another construct.

Everyday experience is encountered, attended to, and rendered thematic and meaningful in terms of a person's unique biographical situation and the provinces he or she shares.[8] Under the natural attitude of the paramount reality of the everyday world, these provinces ordinarily do not conflict. Each province has its own cognitive style according to which experiences in the world are structured and made inter-consistent. We are not equally interested in all the strata of the working world at every moment, nor in all of the other provinces of meaning simultaneously. For instance, a person may be simultaneously a mother, wife, Episcopalian, golfer, and a doctor. While attending to any one of these provinces of meaning (through the selective function of interest), the person will act under the influence of the relevant accent of reality.

As a finite province of meaning, the medical profession is characterized by its own *style*.[9] The physician shares with other doctors a particular mindset or medical vision (a clinical perspective). Physicians suspend doubt in the meaning and validity of their work, its presuppositions, and the knowledge on which they base their work. Their inculcation in method and decision-making serves to make their actions "second nature," and thus appear spontaneous in situations where others would not act. As a result, physicians take on an identity as members of the medical profession. Their relations with other physicians, with patients, and with the public possess a particular tone. Finally, as a member of a finite province of meaning, the doctor is divorced from the time horizon of the everyday world. The physician's sense of time is not the same as it is for patients and their families. He or she defines a problem in light of certain goals of medicine: diagnosis, treatment, and prognosis. Outside of a medical emergency, these goals are accomplished over time. The patient, however, seeks explanation, cure, prediction, and a qualitatively immediate return to the normalcy of daily life. The time it may take for this process to be accomplished is viewed differently by each party. The doctor views his or her time perspective as typical for all people. The patient cannot understand what is taking so long [10].[10]

The public may form a picture of doctors' attitudes and behavior, supporting the expectation that all doctors should act a certain way at all times. However, physicians view themselves and their work from within their own biographical situations and their life plans. Any sense of a common identity results from the profession's style which they incorporate from its tradition and training. Therefore, they view themselves, their colleagues, their relations with the public, and the meaning of their work from a "medical" point of view.

Every province of meaning contains a system of relevances that determine the selective function of a person's interests. Due to the constructs of typicality that characterize the everyday world, the doctor possesses a common sense stock of knowledge of the world and its items (including other persons). However, the social typification of the medical profession incorporates and reflects systems of relevance that establish interests and priorities for its members, and define the particular problems that are worth attending. In medicine, the physician attends or focuses on experience distinctly. Due to their training, physicians are less observers of discrete disease entities than intense readers of signs. He or she is trained to "read" illness essentially as a collection of physical signs and symptoms that describe particular diseases. A physician thematizes the illness as a particular or a typical case of multiple sclerosis or cancer ([274], p. 222). Diagnosis in the medical province represents a "telling" of what these signs reveal. The doctor spends years collecting and studying signs, and this activity basically sets the pattern of his or her observations thereafter. Physicians see and interpret what they are trained to see [53]. Given the medical province of meaning's stock of knowledge and systems of relevance, the doctor adopts a typical focus or way of attending to reality and, therefore, a method of relating to people.

For the physician as a status professional, the medical accent of reality is fundamental. It circumscribes the nature of medical work and determines the tone of relations with other provinces within the world of daily life. The relations between physician and patient become complexly choreographed. Any variation to the expectations of each party becomes unsettling.

The motivation for focusing is related to the physician's place in the familiar world of medicine. For example, in professional practice, habits develop according to which reality is interpreted. Such habits represent a distinct approach to the world and compose the profession's culture ([274], p. 223; [156], pp. 6–7).[11] These habits determine the manner in which an object of experience is rendered thematic and made explicit through the typification process. Thus, given the systems of relevance and the habits that result, the physician operates within an *horizon of meaning*, which in turn becomes a motivation for constituting reality according to the profession's frame of reference. The profession contains its own specific tension of consciousness or horizon of meaning that determines a person's typical behavior or *purpose at hand.*[12]

Under the medical profession's accent of reality, there is a process of self-typification through which physicians define the nature of themselves and their work. The roles, codes, and traits considered valuable by the profession represent a construct of identity which physicians absorb: "I am a doctor; doctors think in this manner. Traditionally, doctors behave in these ways." The physician also is a man or woman, husband or wife, father or mother, friend, and human being – a member of the world of everyday life. The accent of reality can shift from one to another of these areas depending on the attention of the person. Being a physician does not make a person necessarily a better mother, father, friend, or human being. However, in matters of medicine the accent of reality of the medical profession is very strong.

A doctor may come to see herself as the physician-who-is-also-a-mother. Physicians argue that the medical accent, given the nature of their work, functions most effectively in rendering reality meaningful. Thus, the medical accent is very strong one and colors the doctor's *standpoint* in the everyday world. When different provinces' values collide, the doctor's view is strong. It is hard for a doctor to *bracket* or set aside the profession's accent. After all, the doctor "knows" that he or she knows more than the patient. The professional takes for granted the belief that he or she can be trusted with that particular knowledge.

In short, the medical profession represents a finite province of meaning characterized by an accent of reality establishing the habits, relevances, and focuses that are taken for granted in a doctor's work. This accent is to be found in the profession's tradition and knowledge passed on in the training and socialization accompanying membership in the collectivity. The role of knowledge in shaping the medical accent of meaning can be illustrated by bracketing the usual assumptions about the way students are socialized or assimilated into the boundaries of the profession through their education, training, and experiences in medical school [87].

The formal knowledge (the theory, techniques, models, and the "facts" held to be significant by the profession in terms of its work) transmitted in medical school identifies important characteristics of medicine. In medical education, this knowledge may take the shape of an essential "core." Other knowledge becomes optional or peripheral knowledge. Thus, medical knowledge reflects a given order in the larger domain of knowledge and the world of work. Hence, the status of formal knowledge as a cultural artifact, as constructed by a particular province of meaning and taken for granted by its members, remains hidden to the public (and to many doctors, as well) ([5], pp. 235–236).

Several sociological studies of medical education assume a distinctive institutional ethos into which students are assimilated. This assumption portrays the socializing agency as homogeneous and segregated from the outside world [14, 50, 189]. However, there is no "typical" medical school, curriculum, faculty, or student. Medical students do not learn "medicine" in any absolute sense. They do not assimilate a single package of knowledge and skills that are then applied throughout the course of their subsequent careers. There is no ideal medicine that exists universally, absolutely, and independently of everyday practice in clinical and social settings.[13] After all, medical knowledge is organized knowledge – the way in which it is organized and thereby taught depends upon the specialty and the medical model chosen.[14] The accent of reality does not derive from a universal set of data, but it does create a generally typical way of regarding medical work.

Granted these variations, two characteristics of the profession's accent that indelibly mark the physician are an emphasis on "personal judgment" grounded in a "clinical perspective" ([5], pp. 236–237). There has been a tendency to regard these emphases as an individual's psychological adaptations to the conditions of his or her work, as part of the socialization process into the ranks of the profession [131]. An investment in personal judgment and a clinical perspective, however, allows the physician to maintain a position of authority even as knowledge and work conditions change [76]. David Hilfiker tells of a supervising physician who would simply declare, following careful interview, meticulous examination, and thorough laboratory examination, that a patient suffered from "the such-and-such virus," a diagnosis that was incapable of proof or disproof. After several years in practice, Hilfiker realized that the doctor, faced with uncertainty and the pressure to diagnose something, needed to make a definitive diagnosis to protect his image of himself as an expert, knowledgeable and in control ([130], p. 61).

Medical knowledge is the key to power and authority within the field of medicine and of the field within society. It is in the self-interest of the profession to foster the attitude that this power is beyond the ability of laypeople to understand. The emphasis on the physician's personal judgment maintains the importance of each physician's personal knowledge and personal experience as a component of the doctor's authority and professional persona. The prerogative for decision making thereby remains in the doctor's hands. The assumption of a clinical perspective serves to separate the physician from the patient and reinforces the value of the physician's expertise. The emphasis on a clinical perspective determines the way knowledge, techniques, skills are selected, mastered, and applied in the world of working.

Medical education perpetuates and reproduces this view of medicine. The practice of clinical instruction reproduces the characteristics of empiricism and the distinctive approach to disease and doctoring of hospital-centered medicine [4]. The emphasis on medical knowledge separates the profession from the lay public, but it also legitimates a partial version of professional work and interests (e.g., a hospital-centered versus home-centered medical care). Emphasis on knowledge as a primary focus in medicine promotes a consensual view of the legitimacy of that knowledge while masking the social differences it serves to promote (between doctor and patient or doctor and doctor).

The development of typical knowledge and the accent of meaning it generates represents a process of "boundary maintenance" that acts reflexively to distinguish the medical profession from other institutions. The medical profession, as a status profession, and the physician's claim to membership in it, revolves around the knowledge and the skills that it holds to be important because they separate it from other occupations in the world of working. The medical profession presents the knowledge it possesses as the best available knowledge for its work. The outward face of the profession is homogenous; it encourages the public to assume that all doctors share the same knowledge, expertise, and values. However, the dedication to knowledge by the profession promotes the appearance of consensus while actually masking segmental perspectives. The profession's accent, which stresses a dedication to knowledge, is seen as incorporating intraprofessional interests and specialties, as well as reproducing the esoteric expertise that is held to be the preserve of the profession as a whole ([5], p. 239). The fact that medical knowledge is socially constructed and controlled obscures the multifarious nature of knowledge and conflicting claims that exist within the profession.

Thus, the like-mindedness of the medical profession can be elucidated by the application of phenomenological concepts. A discussion of focusing, habits of mind, relevances, finite provinces of meaning, and accents of reality illustrate the manner in which an individual physician actively constructs the meaning of his or her experience from within a particular horizon of meaning. While there is recognition that physicians' conceptualizations differ from their patients' experiences, the difference is often assumed to be simply a matter of different levels of knowledge with the physician's conceptualization being regarded as the more accurate. However, as Toombs points out, a phenomenological analysis of the "reality" or meaning of illness reveals more than a difference in knowledge levels between patient and physician. There is a difference in understanding or interpreting illness, as well [274]. The doctor's understanding is shaped in significant degree by the profession's horizon and accent of meaning. This professional understanding becomes typified as part of the knowledge necessary to the physician's work, but it may not correlate with the patient's experience of illness-as-lived.

The physician adopts the stance of a disinterested observer, armed with knowledge and practical skills, who subscribes to the clinical perspective, and who interprets experience from a particular point of view. Within the medical province, the doctor is the center of his or her own existence while dedicating himself or herself to the welfare and well being of his or her patient. Selfless service or altruistic service is presented as an ideal of professional work. The medical accent of reality includes the value that the physician is one who transcends the intentionalities of the everyday world to be of service to those who need help. This value distinguishes the medical professional attitude. However, it also places doctors in a position of power – the power to help – and allows them the social license to control the limits of their work.

THE INFLUENCE OF THE BIOMEDICAL MODEL ON THE MEDICAL ACCENT OF REALITY

Scientific knowledge, as represented by the biomedical model, is a major component of the physician's definition of "knowledge." This knowledge, and this way of knowing, separates the physician from the general public. I am not arguing that all doctors possess the same knowledge, scientific or otherwise, nor am I engaging in a fashionable critique of the medical model

[68]. However, because of its influence on the profession's accent of meaning, scientific knowledge affects the manner in which physicians typically define their relations with others. Thus, the biomedical model and the behavior it engenders are assumed by doctors in their evaluation of competence in themselves and their colleagues, and affect their relations with patients. The typical patient, or member of the public, ordinarily does not share this perspective. This difference in horizons represents the potential for conflict and misunderstanding, even as it serves to delineate the physician's personal and social identity.

Pellegrino argues that in the twentieth century, medicine stresses clinical observation, careful study of natural history of diseases, and the cultivation of a fund of empirically verifiable data as the basis for its work (and its claims to authority). An "empirical" science describes the particulars of objects in the everyday world and forms generalizations referring to and based on these typifications. Generalizations refer to all occurrences of a type or class of object. Wedded to a scientific and clinical vision, modern medicine is based on the conviction that human illness can be described in physico-chemical and quantified terms ([74]; [221], pp. 11–12). The increasingly close alliance of clinical medicine with physiology, biotechnology, biochemistry, pharmacology, and other biomedical sciences in the last several generations has resulted in great advances in the armaments of clinical medicine. These advances also raise philosophical and theological questions as the alliance shapes the habits, focusing, and relevances of doctors and thus the meaning of "medical care."

Philosophically, science is a "public" enterprise. It endorses a method based on the communal and empirical verifiability of its claims. In another crucial sense, however, science represents a perspective and an investigational method distinct from the attitudes of the everyday world. Its isolation from the world stems from the revolutionary nature of its search for truth and is one of its most distinctive features. By seeking a unified and internally consistent interpretation of meaning of the world and the relations of objects within the world, science requires a fairly strict demarcation from the fuzzy, ad hoc, and heterogeneous meanings of everyday life, in which "reality" is taken for granted, ordinarily unquestioned [265]. Science brackets the natural attitude of the everyday world and incorporates careful and systematic questioning as a basic component of its "work."

In this demarcation or bracketing, the orientation of science lends itself to its own *style*. When applied to medicine, this style (or horizon of meaning) creates: 1) parameters of meaning for each physician; 2) parameters of meaning for the profession itself (of which each doctor is a part whether closely or distantly); and 3) parameters of meaning for the image of the physician presented to clients, government, or the public at large. The alliance of biomedical science (and the scientific horizon, generally) creates a "perspective" not shared by the public or the patient. Science "sets off" physicians and the medical profession from other people and other institutions. This distinction creates certain expectations of medical work on the part of both doctors and patients.

The world of scientific contemplation is quite distinct from the naively experienced, immediately perceived reality of the everyday world ([243], p. 232). While engaged in a scientific project, scientists detach themselves from their biographical situations and adopt the accent of meaning relevant to the scientific province. Relevancy in scientific work may be quite irrelevant in the scientist's daily life and vice versa. By stating the problem at hand the scientist defines relevant considerations and guides the process of inquiry ([243], pp. 37, 249). Doctors attend to their work in the "disinterested attitude" of the scientist. They distance themselves from their own existential situation to achieve objectivity, and establish their decisions on clinical observation and a stock of empirically verifiable data ([275], p. 226).

Such an attitude claims an autonomy and freedom in work. To accomplish the goals established by science's horizon of meaning, the scientist believes he or she requires and deserves some special dispensation from the ordinary demands of everyday life. From within the medical field's horizon of meaning, the physician is working to help the patient and, therefore, deserves the autonomy and authority necessary "to do the job." The pursuit of additional medical knowledge and expertise is for the benefit of present and future patients (thus, the public at large). As medicine moves to an alliance with science, the physician's accent of reality shares this expectation of autonomy and freedom in matters of clinical work or research.[15]

The medical accent of meaning readily adopts a scientific orientation. The phenomenological concepts of a finite province of meaning, focusing, relevances, and habits of mind provide a model for the way medical science is organized, and, thus, constitute the horizon from within which physicians see their "work" and their relations with others. The habits of mind that render experience thematic define a doctor's emphasis in regard to his or her practice. The scientific habit of mind provides a horizon of meaning, a motivation for focusing, and a means of constituting reality quite distinct from other interpretations. According to the medical profession's habits of mind, illness and relations with people are rendered thematic in terms of the "objective," quantifiable data.

Yet the very procedures science maintains also serve to limit and curtail scientific investigation. The "best" scientific writing is practical, precise, orderly, and both usable and disposable. It serves not as literature but as an aid to collegial communication. A "good" diagnosis, a clear medical chart, and assertive treatment establish the reality of a situation by reducing its ambiguity.

However, the everyday world is interesting to the members of the scientific world, whether as an object of study or an object of practice ([265], pp. 71, 74). The world outside of the scientific province of meaning remains the world of paramount reality, even for scientists, and trades more freely in ambiguity. Therefore, it constantly intrudes into the finite province of science in a process that continually threatens either the rigor of the analyst or the faith of the practitioner.

Specifically, scientific knowledge ineluctably derives from questions largely determined socially and culturally rather than by the object of study itself. For instance, there is a general public belief in the scientific basis of medicine precisely because disease and illness are perceived as natural, malignant phenomena. Thus, the knowledge gap between patient and doctor ordinarily is not resented by patients because doctors are seen to use their knowledge and skill against a natural enemy [134]. The demands and interests crystallizing in professional practice constitute one of the mechanisms through which such determinations take place. If the concerns of everyday reality change, as they will over time, the constructs, typicalities, and meaning systems of science and medicine also will change.

Scientific knowledge is always preliminary and is contingent on paradigms of fundamental concepts, assumptions, accepted modes of investigation, and standards of empirical adequacy ([237], p. 49). These paradigms change over time. For example, it was once believed that uroscopy, an ancient diagnostic procedure that involved the visual examination of a patient's urine, provided a viable means to gain knowledge of the patient's illness. Paracelsus challenged this prevailing paradigm by arguing that chemical analysis could reveal signs of disease hidden from visual examination. Although this proposal was criticized and strongly resisted by those practitioners who stood behind traditional, authoritative diagnostic knowledge, over the next few centuries the introduction of chemistry into medicine established the laboratory as an important element of diagnosis ([233], pp. 122–123).

Scientific work leads to its own eventual modification. This revisionist quality does not lead to the abandonment of science and the scientific method. Instead, it reinforces science's authority in the eyes of its adherents.[16] Most practicing physicians take for granted the structures of typicality offered by the biomedical model. Any call for abandoning or revising biomedicine becomes a challenge to the accent of meaning that is the basis for their autonomy and authority, and is resisted.

PROBLEMS OF A SCIENTIFIC APPROACH TO MEDICINE

There is a changing perception among physicians that the biomedical model of illness and an emphasis on curing represent necessary and sufficient conceptions of medicine [105]. However, the taken-for-granted value of "scientific medicine," rooted in its commitment to expertise and technical skill, remains an important component of the medical professional's sense of identity and meaning. As I will show in the next chapter, the way the physician gives and receives or accepts trust is determined by the medical province's horizon of meaning, and, thus, shapes the manner in which he or she relates to others and understands and reacts to moral dilemmas. The climate of suspicion surrounding medicine fosters the illusion of trust within the health care relationship. This climate results from problems caused by the biomedical orientation to medical care.

Due to the demand for theoretical detachment, the scientist's primary role is that of commentator rather than practitioner in the everyday world. Within science, one's audience and judges are solely one's colleagues who, ideally, share a purely disinterested commitment to truth. The everyday world, however, has an interest in work that is both more practical and more immediate ([265], p. 72). In attempting to be a scientist-physician, the doctor is caught between the demands of the scientific orientation within which he or she operates and the immediate concerns of the people whom he or she professes to serve. As we have seen, the set of rules and practices by which the boundaries of scientific medicine take shape have a special form. Since medical work takes place in the world of paramount reality, it must be practical and have practical effects. However, physicians' work involves interactions with others who do not share the profession's accent of reality and horizon of meaning. The goals and values of doctor and patient may not coincide and may lead to struggles over autonomy and authority in medical care.

The medical profession's claim that autonomy and authority are necessary to the work of physicians is understandable, inasmuch as the profession is a social institution interested in controlling the parameters of its work. While maintaining nominal control through licensing, society allows the medical profession great latitude in defining and conducting its "business." The government requires licenses to ensure that professionals are qualified, yet these licenses also allow the medical profession to police its own ranks, to be the judge of its limits of expertise and knowledge. Rights and privileges are granted to physicians by both statute and popular approval in return for specific duties and responsibilities.

If the profession's claim to autonomy and authority is based on expert knowledge gained from the alliance of medicine with science, certain problems result. The activity of "doing science" has no general statutory restraints.[17] Those who work in the natural sciences are not subject to the same types of formal regulations, public restrictions, or controls as the professions. When allied with medicine and the professionals who practice it, biomedical science creates a powerful impetus for the pursuit of knowledge and an accompanying desire to put that knowledge to use ([146], pp. 211–234). The combination of biomedical knowledge and technological proficiency enables physicians to characterize themselves as gatekeepers and controllers of power and expertise. This combination encourages them to feel less restrained by social control mechanisms such as regulation. It also encourages a tendency towards objectivity and exclusivity on the part of the profession in its relations with the public. Physicians take for granted the expectation that great latitude in autonomy and authority is necessary in their work.

Science, however, is only one theoretical province within the world of daily life. Within the boundaries of the scientific horizon of meaning, premium emphasis is placed on working with objective, clear, and precise technical data. It becomes easy to place an emphasis on technical proficiency and argue for value-neutral technical judgment. The clinical perspective, or the concentration on technical matters, helps isolate the physician from the pluralistic concerns of everyday life. And yet medicine, at a fundamental level, involves persons interacting in the "real" time and bodily nature of everyday life. A clinical judgment, e.g., to prescribe Demerol or Darvon for a compound fracture, is partly a technical matter, but the decision is not value-free. Given a similar medical situation, two doctors may prescribe different analgesics depending upon their understanding of human values, attitudes towards pain and suffering, and compassion for the patient ([64], p. 224). Therefore, medicine cannot be reduced to a value-neutral science, although the expert knowledge upon which it builds largely derives from the scientific accent of meaning and the scientific method. Medical work involves the physician with the physical lives, beliefs, expectations, and values of people who occupy the multiple realities of the world of daily life. The physician knows more than a patient about medicine, but biomedical knowledge is not all there is to medicine.[18]

All scientific experience necessarily presupposes pre-scientific experience.[19] A cardiologist is concerned with "heartbeats" only because he or she already knows in a pre-scientific fashion how important the heart is in human life-experience [246]. Each province of meaning within the world of everyday life, e.g., science, medicine, or religion, filters everyday reality through its own accent of meaning. It is from the struggle of "perceptions" that the lifeworld forms a "consensus" regarding the reality of human existence, and in turn, this filtering comes to shape the lifeworld. This process is fundamentally social, slow in evolving, and thus "created" in every act of abstraction.

TRUSTING PHYSICIANS WITH THE FACTS

The modern practitioner possesses knowledge of and access to medical techniques, procedures, and medications that far outstrip the practice of fifty years ago. The marriage of medicine and biomedical science has been a remarkable success. The medical accent of meaning's investment in the scientific method provides physicians with a degree of quantitative and qualitative knowledge of health and illness that is superior to the general public's knowledge. At the same time, the particular characteristics of a clinical perspective or medical vision are rooted in this accent. The organization, control, access, and transmission of medical knowledge is structured by the profession and serves to shape the attitudes of the profession's members. The medical profession becomes the institutional repository of knowledge and skills, and controls admittance to its ranks through a credentialing process, as well as maintaining some degree of control in the licensing of its work [87]. Thus, the cultivation and possession of superior knowledge is a necessary professional trait, but such a trait is acquired (and required) to enable the physician to practice an occupation. Inasmuch as physicians "work" in the everyday world, their knowledge and expertise represent elements of social, economic, and political power.

The medical profession is an example of a finite province of meaning. It reflects or typifies a particular way of focusing, habits of thought, and an accent of reality that are taken for granted by its members and determine their view of "work." The technical/scientific component of medical "work" is a primary determinant of the accent of reality that characterizes the natural attitude of the medical profession. This attitude reinforce's the view that more knowledge and more expertise is a basic "good." Thus, a doctor thinking about bioethics or humanistic medicine remains influenced by a biomedical model taken for granted as constitutive of the knowledge he or she requires to do "good work." Recent attempts to balance the empirical and scientific side of medicine with an emphasis on humanistic medicine shape the physician's understanding of the meaning structures that lie behind his or her work. However, these attempts do not suddenly and thoroughly transform medicine.

Medicine's movement towards a biomedical, scientific base has led to the evolution of a particular accent regarding the meaning of its work and the structure of its relationships with those within and without its boundaries. While their biographical situations and their specialties within the profession vary, doctors see themselves, their colleagues, their patients, and their work from a general perspective created by the cumulative tradition and horizon they share as members of the medical profession. Criticism of a particular medical model or construct of reality is seen in terms the accent of that construct. Such criticisms are "explained away" in terms of the model. When such reductionism fails, a genuine crisis of meaning results until an acceptable accent of reality is restored. The critics of the biomedical model, and the professional behavior it shapes, come to their criticism over and against the prevailing model.

The biomedical model has been a major influence on the physician's accent of meaning and the medical province. Even as criticism has been directed at the effect this model has had in shaping the typifications of the physician-as-professional, the medical profession will not return to its pre-scientific days. First of all, the scientific orientation leads to increased knowledge and expertise. Second, a scientific orientation in its work results in therapeutic tools to enhance caring and curing. Third, the development of a scientific orientation within the profession has led to increased power and authority for physicians. People trust doctors, at least in part, because there is a scientific basis to medicine. They assume (and want to assume) that doctors know what to do (and know what they are doing). The profession's investment in knowledge and expertise is designed to enhance the public's trusting attitude. It also fosters the granting of power and autonomy to the medical profession (in a mandate for research and the exercise of knowledge-as-power). The profession's commitment to scientific knowledge is fundamental to its work and its identity as a province of meaning.

According to phenomenological theory, the world of working "is the archetype of our experience of reality. All the other provinces of meaning may be considered as its modifications" ([243], p. 233). The province of science is more theoretical. When allied with medicine, however, it has practical applications and effects in the everyday world. Therein lie the roots of the public's present crisis of confidence. The strength of the accent of the biomedical province has created strong organization, attitude, and purpose among physicians. The result is a great advance in medical "formal knowledge" and the expertise based upon it, and the taken-for-granted claims by physicians to professional autonomy and authority.

In the next chapter, I argue that the professional and social commitment to knowledge by physicians, unfortunately, also fosters an uneasiness, a suspicion, on the part of the public in its contact with the medical province. The basic orientation in medicine towards biomedical science nourishes an inverse psychological and social relationship between doctor and patient. The greater the drive towards knowledge, the more power and authority the profession exhibits. In spite of additional knowledge, medical technology, and miracle cures, the public feels less satisfied and more uncertain in relation to physicians.

Even as the clinical perspective or accent of meaning leads to medical advances, medicine cannot resolve the questions its success creates. The very success of biomedical science produces a concomitant public desire to reassert social control over the medical profession. Since the public cannot truly compete on the level of professional knowledge and expertise, there arises a natural desire to use trust as a form of social control. This state of affairs leads to "shock" and dis-ease as suspicion and an erosion of trust grows within the health care relationship between physician and patient. This development reflexively affects the physician's sense of the fiduciary focus that gives meaning to his or her work.

NOTES

[1] This tradition commonly is believed to originate in the Hippocratic Corpus, especially in the Hippocratic Oath, which stresses *primum non nocere*, and indicates that the patient's good is to be placed before the physician's self-interest. A few other historical examples of the obligation of physicians to serve the public or patient are Perceival's *Medical Ethics* (1803) and the American Medical Association's first code of ethics (1847); see ([232], pp. 18–34, 25–33).

[2] The description of a profession as a "like-minded" group of people is taken from ([96], pp. 59–60).

[3] The term "Lebenswelt" refers to the pre-scientific and pre-theoretical levels of the world of everyday life. For Husserl, the life-world is the world of everyday experiences as they present themselves to the individual in his "natural stance," the world as experienced and made meaningful in consciousness [139].

[4] Few practicing physicians find the biomedical model fully satisfactory in caring for patients. By emphasizing the scientific province and the accent which this focus places upon medical practice, the biomedical model does not encourage a doctor to deal with his or her own concerns as a person dealing with other people in the world of daily life ([246]; [40], pp. 20–23; and [274], p. 235). However, medical education and curriculum established by academic or research oriented professionals still is influenced largely by the model; see Carlton's discussion of clinical inculcation ([37], pp. 65–83).

[5] Among the many types Gurvitch mentions, predominant ones penetrate all the others in some form. For instance, philosophical knowledge and perceptual knowledge are ingredients in scientific knowledge ([111], pp. 23ff). Gurvitch does not refer to theological knowledge although he refers to a mystical form of knowledge as the supernatural element that can originate in judgment or experience. To him this form usually falls before other forms such as the rational and the empirical (p. 38).

[6] Whatever tasks a doctor performs, whether routine or unusual, are measured against the standard of what Gordon Horobin calls "the great save." One irony of medical work is that the most dramatic evidence of work performance, i.e., the saving of a life, is brought about through the use of esoteric skills for which the physician has trained and yet, by and large for many doctors, remains unused and untested [134].

[7] Schutz refers to the "finite provinces of meaning" since it is the meaning of experiences and not the ontological structure of objects that constitute reality (see [243], pp. 3–47).

[8] William James noted that the manner in which an object is experienced is correlative to the way in which an individual explicitly attends to it. The meaning of the objects experienced will change as the attentional focus varies. "Each world whilst it is attended to is real after its own fashion; only the reality lapses with the attention" ([143], p. 293). Ultimately what the individual attends to depends upon his or her biographical situation and upon the complicated texture of choices, decisions, and projects that make up his or her life plan. Experience is encountered, attended to, and rendered thematic in terms of the individual's unique status: in light of his or her special interests, motives, desires, religious and ideological commitments (see [139], p. 108).

[9] According to Schutz, in any province there are six characteristics of style: a specific tension of consciousness, a specific epoché, a prevalent form of spontaneity, a specific form of experiencing one's self, a specific form of sociality, and a specific time-perspective ([243], pp. 207–259).

[10] Zussman argues that physicians would have a hard time conducting themselves according to medical ethics even if they wanted to, because the realities of clinical decision making differ from how decisions are talked about in the ethics literature. Ethicists often postulate a moment of decision, but for physicians decisions are not an event but a process ([296], p. 159). What is done is not so much decided at a point in time as it is made inevitable by numerous smaller decisions that have preceded it (see [81]).

[11] "Habitus" is a distinctive mode of perception, of thinking, of appreciation, and of action associated with any collectivity. The habitus defines the "taste" of a group – its character and its taken-for-granted view of the world. Applying this idea to the medical profession, the habitus of medicine is the accent of meaning adopted and endorsed by the profession that establishes its "style" of acting, thinking, and valuing (see [5], p. 237).

[12] The "problem" or "purpose at hand" is a definition of what a person considers relevant in a situation ([243], p. 249).

[13] All knowledge transmitted through education is in a sense arbitrary in that there exists no absolute, pre-given corpus of knowledge that self-evidently presents itself as a curriculum, inherently endowed with order and sequential organization, acceptable to all experts in any field. A curriculum is a device of cultural imposition whereby knowledge is classified and combined. There is no ideal medicine "out there" to which the curriculum corresponds as a mere reflection or copy ([5], p. 235). Yet medical students do learn some medicine – or some version of it – through their absorption into the accent of meaning, the habitus, the

relevances, habits, horizon of meaning, and the knowledge and skills that characterize the profession during a particular historical period.

[14] It is important to note that there is professional segmentation within a medical school. A faculty claims "professional autonomy" in the conduct of its work, and exercises that autonomy in light of segmented interests and allegiances. A medical school's curriculum is organized and divided according to specialized bodies of knowledge. It reflects the specialized segment's position in the order of knowledge and social relations ([29], p. 70). Therefore, it must be recognized that medical knowledge varies among physicians. For some, it represents technical expertise; for others, it is bedside manner or knowing how to manipulate the health care system. All share knowledge relating to medical matters in the broad sense and in this possess more knowledge than the patient-public.

[15] Under the medical province's horizon, the physician feels that he or she "deserves" these freedoms. He or she "expects" and "requires" autonomy and authority – they are "necessary" to medical work. Without them, this work would suffer (see [104]). Those who question this taken-for-granted attitude are perceived as outsiders, as threats to "professional" work.

[16] Kuhn's discussion of the process of normal science shows that the emergence of anomalies does not erase a scientist's allegiance, trust, or faith in the scientific method, but rather reinforce his or her commitment to the pursuit of a more adequate paradigm ([160], pp. 66–110).

[17] Although controlled by the standards of the scientific community, scientists generally are not licensed or controlled by the public. This "autonomy" influences the physician's claim to authority (see [278]). Of course, the organizations through which scientific research is applied may be regulated. Heavy government financial investments since World War Two have led to some informal controls. As medical science moves into the area of genetics, there are signs of increasing government concern. However, scientific investigation itself has generally occurred without stringent government controls. The feeling has been that scientific knowledge rebounds to the benefit of society, and restrictions would limit the possibilities of scientific gains.

[18] Some physicians may respond to the demands of ethics by explaining their decisions on the basis of technique rather than values, thus making "principles appear to rest on exclusively technical judgments" ([296], p. 124). The boundaries between the technical and the ethical are, at least in part, social constructions. They are social constructions because technical judgments inherently are probabilistic (p. 151). It is a social judgment what percentage of success justifies some medical action [81]. Again, the physician's knowledge must encompass more than pure technical or biomedical scientific knowledge. The doctor must know his or her patient's wishes, feelings, and desires, and the way in which the physician's own values play a role in treatment decisions.

[19] Natanson argues that the world of immediate experience has a precedence over the derivative world of science. The world is experienced first in its immediacy, and only upon reflection do we thematize experience in theoretical terms and scientific constructs ([198], p. 95). Husserl states that the lifeworld is the "foundation of meaning" for science (as it is for all of human existence), but this basis of meaning has been forgotten because of the "pervasive spirit of abstraction" that dominates science ([139], p. 317).

CHAPTER 4

THE FIDUCIARY FOCUS OF THE MEDICAL PROVINCE OF MEANING

A fiduciary commitment to a client characterizes the physician's professional status and is the second focus that marks the medical profession's accent of meaning. The fiduciary nature of the relationship is a key element defining the boundaries of the world in which the physician operates, the world-map as drawn from the medical profession's perspective. By asking a physician for help, the patient enters this world, this finite province of meaning, yet the patient retains a "practical" interest in the world of everyday life. He or she is not interested in theoretical matters, e.g., the philosophy of medicine or the nuances of molecular biology. The patient's paramount interest is to "get well," to return to his or her everyday reality for which illness is an intrusion or interruption ([295], pp. 53–91; [274]).

The patient expects the physician to share this practical goal. However, theoretical reflection and practical problem-solving are separate activities. The patient can concentrate on the practical because he or she "knows" or believes physicians have these theoretical matters as their practical concerns. After all, the social distribution of knowledge is predicated on the assumption that no one member of society need know all of society's knowledge to interact with each other ([186], pp. 104–105). To function in everyday life, each must rely on the fact that some people have some knowledge of matters within the world while others have different knowledge, and that in their interaction, some mutual benefit will accrue. In phenomenological language, this confidence is represented by the natural attitude of everyday life, in which, until proven otherwise, people assume confidence in the reality of the social world. The event of social interaction possesses a depth of meaning for both participants that represents more than either one can say; both know in common what cannot be said in so many words ([93], p. 561).

However, a shock to the natural attitude occurs when one's taken-for-granted confidence in "reality" is upset, and typical responses fail to restore the epoché of the natural attitude. Ordinarily a person finds a way to reconcile the anomaly in his or her horizon of meaning so that confidence (the familiarity of the natural attitude) is restored. During the period of anxiety or uncertainty, a person perceives the risk and dependency inherent in social interaction. Rather than trusting that the intentions and disposition of others are mutually beneficial and not merely advantageous to the more powerful person, he or she realizes the fragility of relationships with others and becomes protective of self-interest.

As a factor in social cohesiveness, as the condition on which rests a certain naivete of interaction, trust acts to maintain the boundaries of social relations. The term possesses moral connotations: If one trusts another person, he or she expects that person to act in predictable ways. If someone is trustworthy, that person is granted a license of power in interactions with another who may be of diminished capacity. The expectation of trust and trustworthiness represents more than mere social and psychological confidence. It becomes highly prized as a valuable and necessary feature of human relations. In a relationship as fraught with emotional and power-laden issues as the medical one, the fiduciary component achieves a depth of meaning that is fundamental in its implications. This fiducial quality assumes a boundary-forming (a protean) dimension. Physicians' self-identity and self-awareness, as well as their expectations of their work and of their patients' response of their work, are shaped by regard for trust and trustworthiness. It influences their expectations of themselves (as technically competent professionals), their profession (its tradition of service), and of the patients they "serve" (as people who, in turn, have obligations to doctors).

The medical profession's claim to trustworthiness rests on the grounds that physicians provide explanation, security, and confidence, a sense of protection to those in need of comfort and consolation. To earn the public's confidence, the doctor must give the impression of possessing expertise and knowledge. However, in order to earn the trust of patients, the physician must possess and demonstrate an ability to be relied upon beyond the minimal levels for psychological confidence. The non-expert expects a great deal of physicians, thereby granting the doctor the duty, task, or charge imposed in faith and confidence that his or her skills will be used or cared for in the interests of others. Here, trust is an aspect of the virtue of beneficence governing the medical relationship, one of the moral commitments that transforms the physician's subjective beliefs and convictions into the "professional" character ([181], pp. 56–57).

THE PROFESSION'S FIDUCIARY COMMITMENT

In return for wide latitude in controlling the limits of its work and the knowledge that supports its autonomy, the medical profession accepts the public's expectation of trustworthiness, incorporating the fiduciary dimension into its own sense of identity. Therefore, doctors endorse and reflect certain traits that support and reflect this claim. Those traits and actions that undermine the claim to trustworthiness are disavowed. However, that which makes for a "trustworthy" doctor may mean one thing to some doctors, something else to others, and something altogether different to a patient [275].

If there seems to be a decline in the public's trust in social institutions such as medicine, what is meant by trust must be carefully considered. The image of the medical doctor as an autonomous authority figure dedicated to the service of humanity is changing [124]. Since medicine involves primarily a social relationship between physician and patient, trust must embrace at least two different meanings when the term is applied to the relationship between a member of a profession and the public. The first meaning links trust with the expectation of technically competent role performance. Patients expect or want confidently to believe that a physician knows his or her business, and thereby is technically competent. The second sense of trust, however, is the expectation that some persons in social relationships have the moral obligation and responsibility to demonstrate a special concern for others' interests above their own.[1] In this latter sense, patients trust that their physicians care about them and have their best interests at heart. In this way patients place a great deal of faith in their doctors and, beyond the individual, in the profession. Because of the public's trust in the profession's technical and fiduciary reliability, the patient places himself or herself in the physician's hands.

Within such a relationship, there is a mutual confidence between or among the parties. Each person is considered to be dependable because each is given and assumes the responsibility to be trustworthy. The meaning of "fiduciary" as a trustworthy person or as a condition of trust has social, psychological, phenomenological, and moral connotations. These connotations allow for social interaction by serving to minimize uncertainty for the everyday world and for the medical accent of meaning that gives definition to the physician and his or her work. Without trust, the everyday social life that we take for granted is simply not possible [168].

Therefore, the difficulty in defining trust illustrates the highly personal and social nature of the concept. Words contain an archeological stratigraphy representing the continuing evolution of meaning and usage. The connotation of trust changes over time and from culture to culture. For example, the Germanic stratum of the English word "trust" stresses the active voice of *ich*; it is personal and transitive. Latin-based constructions yield agency to abstractions, masking the speaker's voice in passive or intransitive constructions. Modern English frequently collapses the distinction between persons and things; trust can refer to a person, an idea, or an object. Linguistically, modern English usage has lost the emphasis on the personal nature of relationships. Nothing stands between an individual subject and the depersonalized, abstract world. Contemporary language exposes us to anonymous phenomena like "health care" or "the profession" which remove from sight the intermediate pattern of human interdependence that gave rise to the words we use. In the postmodern view, words do not stand still, and any attempt to define them finally will fail. In its linguistic origin, however, trust refers to an experience explicitly found in concrete human relations.

Also, there are fluctuations between the notion that trust is something specific within a society (dealing with relations between particular persons) and a broader notion that trust is coextensive with the very existence of the social order (that trust functions as the cohesive factor fundamental to any discussion of social order). This ambiguity may be an inherent characteristic of everyday reality. Given the diversity of social order, how can trust be said to establish social order? Introducing the social nature of trust carries the discussion to the phenomenological dimension of trust.

FIDUCIARY COMMITMENT AND THE MEDICAL PROFESSION'S NATURAL ATTITUDE

The fiduciary attitude generally is characterized by the expectation that some in our social relationships have a moral obligation and responsibility to demonstrate a special concern for others' interests above their own ([7, 168]; [128], pp. 1–22; [249]).[2] I have argued that a status profession such as medicine is entrusted with power and authority inasmuch as the public believes that the profession is acting in a client's best interests. Doctors certainly have other interests and roles that may conflict with a fiduciary attitude (they belong to other finite provinces of meaning besides the medical one), but the profession's accent of meaning includes the moral duties of beneficence and non-maleficence to the sick.[3]

If a fiduciary component is ascribed to the professional identity, the physician becomes a trustee of the patient, a person whose work rests upon public confidence for its value. Faced with a situation of illness in which the ordinary boundaries of their everyday world are disrupted, patients enter into a special relation with their physicians. To maintain the expectation, hope, confidence, and faith that the doctor knows what he or she is doing and will act in his or her interests, the patient is willing to suspend doubt in the profession's technical competence and fiducial commitment. Trustworthiness becomes a characteristic of the "good" doctor and is expected of any physician. The physician becomes a figure whose authority and trustworthiness is accepted by others to a remarkable degree. Therefore, trust has a particular significance within the medical province of meaning.

Within the medical accent of meaning, the doctor assumes that others share similar preconceptions, values, knowledge, sense of time, and language. A physician studies the life-world from the perspective of the profession and his or her specialty within it, with the knowledge he or she has learned and gained from experiences of similar situations. A doctor assumes that other health care professionals share, more or less, a similar understanding. This clinical stance represents an accent which accepts a certain way of seeing the world, and which determines the meaningful structures within which the doctor operates. This accent involves the assumption of a *reciprocity of perspectives* and the *typicality* of others in the world ([243], pp. 143–203).

Under the reciprocity of perspectives, I assume (and assume that you assume) that the various objects and events in my world are as accessible to others, in general, as they are to me, or can become accessible to others ([292], pp. 84–86). Thus, for practical purposes, I assume that the matters about which I am talking also can be understood by you, within certain limits. I also assume that differences in our respective stocks of knowledge and biographical situations can be ignored in general for our specific purposes in the task at hand. Ordinarily, in other words, we assume that objects and events have a similar meaning to us both.

Further, only a small part of my knowledge is unique to me. After all, for the most part, what I know is handed to me by others. Such knowledge includes not only ideas and values, but also how to define and interpret the world, how to find my way in the world, how to think, and to act. This knowledge is contained in the typifications prevailing in any group sharing the same "common language." Such language reveals the prevailing texture of "definitions of situations" and typifications in any specific group ([293], p. 39).

Physicians experience, interpret, and act in and on the world with the assumption that not merely are there "others," but that these others are different types – strangers, friends, the sick, the healthy, patients, and other doctors. To each of these types are correlated expectations that are taken-for-granted typifications in the doctor's daily life.[4] The physician expects particular expressions of thought or action from *typical* individuals (e.g., a hospital accountant will be unable to understand the necessity for additional lab tests; a person with a rotator cuff injury typically will experience pain in the shoulder). Since not only the physician but everyone belongs to the life-world, and therefore is endowed with the same activities of consciousness as he or she is, doctors assume that other people in similar situations also have the same expectations of these types.

In sum, a reciprocity of perspectives and typicality are tacit features of social knowledge. An unspoken and common understanding of "the world" is assumed to be shared until otherwise apparent. These idealizations build a mutual sense of reliance and normality into everyday interactions. Personal disparities in knowledge are less threatening in these interactions (it is accepted that some know things that others do not). Finally, people take the world for granted by assuming that the real world exists independently of one's knowledge of the world, and that these two are in direct correspondence (i.e., that knowledge and world mirror each other). The assumption of a reciprocity of perspectives and typicality allows people to rely on their perceptions of others in social reality.

In the doctor-patient relationship, and in health care matters generally, these idealizations are complex. Doctors, other health care workers, and patients do not share a complete interchangeability of standpoints, nor can they ignore the differences in their respective biographical situations. The reciprocity of perspectives is not absolute. I can accept the possibility that objects and events mean something different to others, since our biographically determined situations, with their respective relevancy-systems and other contents, may differ. For instance, an ambulance siren is a signal to pull over both to the layperson and the physician, but the physician reacts to the siren in a certain way due to his or her biographically determined situation and the particular meaning that emergency systems elicit.[5] The call, "Is there a doctor in the house?" is a commonly understood call for help, but a physician's response is different than the response of a lawyer. The patient's biographical situation is precisely what has brought him or her to the physician. It is their difference in biographical situations and stocks of knowledge that creates the relationship.

In addition, the two people do not share a fully common language since the medical profession's language is quite specialized. There is common ground in the everyday language of a culture, but the medical province of meaning has a language all its own. It has to be "translated" if the patient is to understand. Such translation is necessary if authentic sociality is to be present. However, by taking for granted that the language of the medical province has common meanings for patients, or by refusing to take time to translate adequately, the physician's power is reinforced. For instance, "gomers" is an expression used by resident doctors to characterize a type of patient [254]. The expression reflects the doctor's view of that type of patient, and defines the way in which a doctor may respond to that person.

However, given the reciprocity of perspectives outlined above, the doctor often assumes that the patient can understand him, and can "check the doctor out," confirming for himself what the doctor says – despite their various differences. Of course, the patient can question the doctor or obtain a second opinion. But part of the taken for granted perception of doctors, and of patients by doctors, is that doctors know what they are doing and that patients should trust them. To ask for a second opinion or to question the doctor is to suggest that we do not trust them fully. Doctors assume a reciprocity of perspective because they believe that they do know what they are doing and that patients should trust them.

Can Physicians Know Patients' Best Interests?

As we have seen, everyday knowledge of the world is knowledge of typicalities, the world is experienced as commonly shared, and this knowledge is taken for granted ([290], pp. 56–57; [227], p. 9). Within the standpoint of the natural attitude, which assumes that the everyday world is as it appears to be typically, the individual is not motivated to question the meaningful structures of his life-world. His or her interest is a practical one – his task to live in, rather than to make a study of, the life-world. Ordinarily we suspend doubt that the world and its objects can be other than they appear. The patient assumes that the physician is trustworthy and suspends doubt in the professional's expertise, even in the face of contrary evidence. The alternative, that the physician is untrustworthy and "incompetent," is too unsettling. It introduces risk and anxiety into the already uncertain situation of illness.

Due to its accent of meaning, the profession takes its own expertise and trustworthiness for granted to reduce the uncertainty inherent in medical work ([149], p. 166). The profession's fiduciary commitment represents a defense against the awareness or acknowledgment of uncertainty. Renée Fox argues that there are basic types of uncertainty for medical professionals. The first results from incomplete or imperfect mastery of available knowledge. No one can have at his or her command all the skills and knowledge of medical lore. The second depends upon limitations in current medical knowledge. There are innumerable questions that no physician, however well trained, can answer. A third source of uncertainty derives from the first two. This uncertainty consists of the difficulty in distinguishing between personal ignorance or ineptitude and the limitations of present medical knowledge ([76], pp. 208–209). The physician must confront and come to grips with these types of uncertainty. The doctor who begins to question his or her own knowledge and abilities can be overwhelmed by doubt and second-guessing.[6]

Casting for coherent forms is an interpretive procedure performed in all knowledge systems. Medical science, clinical technique, and professional tradition provide the doctor with an horizon of meaning and an interpretation of events. This interpretive procedure is called "searching for a normal form" ([186], p. 103). The doctor must transform the swirl of stimuli in every medical situation into a meaningful whole. The physician searches for and selects features of the world that can be placed into familiar schema.

Confronted by swirling data and driven by the need to impose order and explanation, the doctor looks for the presence of a causative agent, though its specific makeup may be absent and unknown. The assumed presence of an objectively verifiable disease entity enables the physician to look for features that help him or her identify the patient's condition. Unclear data is set aside or held in abeyance while the physician seeks clarification or confirmation of other data. He or she assumes that subsequent events will clarify any present ambiguity.[7]

Although a symptom may not have precise meaning when initially apprehended, it has some meaning – it is assumed to be a sign or a trace of disease or physical disturbance. Specific meaning becomes clear with subsequent events and investigation. For example, the expectation of medical tests is that they will provide clarification and certainty, to the point of formulating therapy. In retrospect, with the formulation of diagnosis or following treatment, a doctor develops an explanatory history of the patient's condition. This retrospective "filling in" is an interpretive process. Thus, doctors will acknowledge medicine's uncertainty once its presence is forced into their conscious awareness, yet at same time they will continue to conduct their practice as if uncertainty did not exist. Of course, any current understanding is also subject to subsequent reinterpretation. The meaning of an object, event, or utterance is "prospective." Subsequent events may alter the "normal form" assigned to it ([186], p. 103).[8]

Doctors, even in the face of the uncertainty inherent in their work, take it for granted that their skills, knowledge, experience, and dedication to the well being of their patients elevate them to trustworthy status. Thus, trustworthiness is a key component of the accent of meaning that serves as a touchstone for the physician's sense of reality. At the same time, however, such status is also due to the trust that doctors are given by the public. The public accepts the assumption of a reciprocity of perspectives, assuming that physicians do share similar goals and expectations with patients. This public expectation largely exists because the doctor believes in the power and efficacy of expert knowledge. Claiming professional expertise and beneficence, the physician accepts the public's expectation of reliability. For the profession, the careful cultivation of knowledge and the acquisition of clinical expertise and perspective enhances the care of the patient, which further anchors public trust. Simultaneously, the commitment to service provides a rationale as well as a sanction for the cultivation and control of "powerful" knowledge.[9]

The profession's dedication to the obligations of service serves a variety of purposes. First, as part of the medical accent of meaning, trustworthiness is elemental to the physician's self-understanding. A physician's self-identity comes from membership in a profession or group that represents the virtue of trustworthiness. The physician sees himself or herself as trustworthy and is identified as trustworthy by colleagues and patients. The fiduciary aspect of professional identity acquires a moral tone [225, 295]. Doctors profess to be trustworthy, because it is expected of them. As long as the profession's fiducial claims serve to satisfy people's expectations, they will suspend doubt in physicians' intentions and dispositions.[10] As long as this epoché of mistrust holds, the physician can have trust or confidence that his or her beliefs and actions are meaningful. Clearly, if a profession owes its status in social life to its fiduciary or fiducial nature, trust is an essential ingredient in "limiting" or defining its claims to authority, autonomy, and power. It is this latter sense of trust that raises the issue of trust as a form of social control.

Second, a dedication to fiduciary relations serves as a form of social control on the part of a profession seeking power and prestige in its work. If the public can be convinced that the medical doctor possesses knowledge and skill, and can be trusted to act in the public's best interest, the profession offers itself as a benevolent alternative to its competition. A cycle of expectations is established: the more the profession dedicates itself to a fiduciary expertise, the more society will accept its claims and the more reliant the public will become upon it. The public expects, or trusts, that the physician is trustworthy and dedicated to service. This trust allows the professional to pursue his or her work while enjoying high social status. In turn, public expectations compel doctors to feel an allegiance to the profession's ongoing tradition of service.

This professional attitude has contributed to a faith in and a commitment to medical science and an endorsement of the promise of success. Physicians place their confidence in science, in their medical training, and in their experience because these generally produce results, as calculated from within the boundary of the medical province of meaning. In return, physicians ask that public give them the autonomy to do their work, claiming that this approach will lead to the best outcome for the patient.

In recent years medical science has produce some remarkable successes, yet there is a sense of public uneasiness with the medical profession. There have been accusations that physicians' power and authority are out of hand, and that limitations must be imposed to restore "good" medicine. Given the accent of the medical province, that expert knowledge and trustworthiness are foundational to their work, it is hard for physicians to understand or accept these calls to redefine the limits of professional power. This professional accent is taken so much for granted that doctors become upset if their expertise and trustworthiness is questioned. The use of distrust to create an acceptable balance of power between the profession and the public creates shock among physicians, and leads to suspicion and distrust in turn. When such a social relationship becomes ambiguous, participants will work to restore a sense of control.

THE BREAKDOWN OF TRUST IN THE FIDUCIAL RELATIONSHIP

The medical profession's authority in society appears strong. Polls show that the profession is rated highly in terms of trust and social prestige [19]. If the profession is held in high regard, why has there been much debate recently about the nature and legitimacy of professional power and its role in relations with clients?

One possible explanation for the debate is that trust in all professions generally is declining. Trust in medicine remains high only in relation to trust in other professions or occupations. Another possibility is that people are forced to trust physicians, given the alternative, yet are uneasy in doing so. Faced with the vulnerability created by illness and neediness, it is difficult for people to question the physician's role, authority, and fiduciary tradition. Finally, it may be that the nature of the social matrix from which trust arises is undergoing revision, and consequently, the public's definition of trust is changing. I believe that concern with the nature of professional power and identity results, at least in part, from the influence of the scientific accent of meaning and its effect on the medical province of meaning. As this influence has shaped in a reciprocal fashion the doctor's work and his or her approach to that work, a misapprehension of the nature and meaning of trust has developed on the part of doctor and patient ([125], pp. 111–112).

It is not surprising that trust is a fragile social condition. Socially extensive and enduring trust is not an easy condition to create. There can be individual motives that aim to accomplish an end other than the individual's strictly selfish interests. It is not unreasonable to suppose that such motives exist at least some of the time. However, these motives cannot be expected to provide for more extensive and properly social trust (such as the fiducial claim made by the medical profession in regards to its commitment to its clients).

Four assumptions must be met if social trust is to exist. We must assume: 1) that people know what each other's motives actually are; 2) that they know that the other knows; 3) that this knowledge is not too expensive or difficult to obtain and maintain; and 4) that the outcomes of any course of action are not too difficult to achieve. Unfortunately, especially given the predilection within the scientific attitude towards clear and univocal perception, these four conditions imply that interpersonal trust is easily obtained ([17], pp. 6–63). Although the medical profession is dedicated to knowledge and expertise as a means of controlling its work, and access to social prestige and power serves to reinforce both professional and public expectations that social trust is easily obtained, such an assumption does not form a firm foundation for authentic or deep seated fiduciary relations between doctor and patient.

In a discussion of the taken-for-granted value of trust as part of the medical province of meaning, a review of the rationalization and monopolization process of professionalization (by means of education, expertise, and codes of ethics) shows the complex structures of power and authority that link the profession of medicine and the public it professes to serve ([190], pp. 1–23; [68], pp. 129–135; [262]). The profession's efforts to achieve a monopoly regarding its formal knowledge and control of its work has moved it toward a close relation with science and the biomedical model of health and illness. Science's promise to wipe out disease, added to the profession's vow to fiduciary service, has contributed to the public's willingness to turn away from other healers and allow allopaths to take charge of the nation's health needs ([149], p. 40). The subsequent development of medical science has resulted in increased specialization and growth in knowledge and resultant biotechnology.

Coupled with a faith in the authority of the medical credential, scientific medicine's success has created unrealistic expectations on the part of the public. This success has led to promises that cannot be kept, and eventual discouragement and backlash by the people whom the physician professes to serve. The medical profession is responsible, in part, for encouraging this expectation since the public's trust necessarily accompanies the profession's efforts to ensure, maintain, and further the expertise that marks its power and identity as a profession ([18], pp. 52ff; [175], pp. 65–67).

The status of medical knowledge at any point is a historical observation about the practice of medicine in particular time and place. In the period since World War Two, the biomedical model has prevailed over alternative paradigms through attack and defense ([277], pp. 41–45). The uncovering of "typifications" by a phenomenologically based sociology of knowledge supports the biomedical model's acceptance and staying power. The search for "normal" forms, the effort to establish categories of disease states, and the focus on predictability and control of medical testing and experimentation create a preference for epistemological objectification. As a result, particular values of relevant physical magnitudes and parameters become the significant focus of medical attention, and so earn the title of "facts." This focus lends itself to quantifications to settle disputes or differences in diagnosis and treatment. The facts used to settle differences in opinions are themselves based in theory grounded in empirical study.

Under the general impetus of an age in which the accent of science largely dominates, facts are taken for granted as arbiters of certainty in the everyday world. To tell scientists or physicians that their model and the knowledge it produces is "only" of a certain form accomplishes little since it does not alter the experiential validity of that knowledge and the biomedical model's success. Physicians assume the validity of the scientific accent of meaning since it demonstrates its power daily. Problems or objections are resolved within the biomedical province of meaning as a course of "normal science."[11] This taken-for-granted validity is hard to modify ([186], p. 221).

The careful control of medical knowledge represented by the process of scientific investigation limits the influence of those outside the profession and enables physicians to control their work. Doctors tend to seek clarity and certainty in diagnosis and treatment because their credibility and authority are at stake. For this reason, physicians rarely criticize or challenge colleagues and certainly not in public. By presenting a solid front, they want to appear competent and knowledgeable.

I am not arguing that doctors as a group are locked into an absolutist biomedical model that treats the scientific method as the ultimate authority for knowledge. First, any community of scholars constructs findings by comparing them to standards that exist at that time. Propositions are moves in the scientific game; they are not direct reflections of nature. Second, scholars do not merely match observations with propositions. They decide truth through discussion, argument, and practical human activities. An organized consensus, reached in accordance with established procedural rules, decides what is and is not warranted as knowledge. Thus, "good reasons" may be offered about why an observation validates or invalidates any particular proposition. Decisions are based on the scientific situation in effect at the particular time. Of course, the scientific situation changes with time. For instance, leeching is no longer accepted as a valid medical proposition. Truths of science (and medicine) are argued and determined in praxis; they are not revealed truths ([186], p. 226).

However, within the medical accent of meaning shaped by the biomedical model, traditional patterns of relationships slowly become reorganized. The demand for clarity and objectivity determines the way in which reality is presented to people (it provides the meaning by which they interpret lived experience), and shapes the way in which they re-present themselves to reality. People begin rearranging the patterns of their lives in light of an increasingly technologically oriented world in which choices continue to proliferate.[12] A demand for additional technology inevitably follows. People believe that if they accumulate further knowledge about an issue they will understand it conclusively, and thereby be able to choose correctly in any circumstance. Unfortunately, each conclusion, each technological development, creates new questions and the need for more knowledge. This phenomenon exacerbates an anxiety or need to know more, to do more research, or to solve the new problem. It has created the specialization of medicine.[13]

The practical effect of this paradox of typification and modification creates an expectation for successful outcome. Due to its very success, medicine is pressured to pursue the scientific direction, which results in more knowledge, more drugs, more technology, more specialists, and some measure of success. The limits of what is possible keep changing, yet limits remain. The public's expectations remain unfulfilled.

Scientific meaning must remain within the particular province or paradigm that determines the meaningful boundaries of the theory or procedure. The universe of science represents a finite province of meaning that is quite distinct from the naively experienced, immediately perceived reality of everyday life. Within its own boundaries the scientific view possesses its own sense of shared activity, language, and meaning based upon a confident evaluation of the authority of reason ([243], pp. 228–229).

As the scientific model became prominent in the last century, it began to shape the social structures of meaning that today typify the medical profession. The orientation of objectivity over subjectivity, of facts over feelings, became a primary value in health care. Crisis medicine, with its heavy technological investment, was the taken-for-granted, most easily followed approach to health care. Preventive medicine was viewed as an "alternative." As a result, the perception persists that a sense of the human element is submerged in a technological orientation to health care.

As science shapes the natural attitude of the everyday world, its boundaries have begun to overlap with other provinces of meaning such as law, religion, and philosophy. However, science cannot provide meaning which cuts across multiple realities. It can give the knowledge of how to do something, but cannot answer the question, what shall we do? ([241], p. 79). The scientific perspective provides knowledge about the world, but it cannot provide a final perspective. As the scientific orientation dominates the medical profession, it carries with it the seeds of the current crisis of trust: the desire and the drive for more knowledge serves to emphasize choices rather than to resolve them.[14]

From the expert's point of view, one consequence of this proliferation of medical knowledge is the inability of non-professionals to make meaningful choices without the aid of experts. A specialist is required to understand a particular medical technique.[15] Because the expert's knowledge is greater than that of the uninformed person, an unequal relationship can develop. The patient becomes dependent on the information given to him or her by the professional. He must trust his doctor's knowledge, and hope that the doctor has the patient's interests at heart.

Unfortunately, as with any situation in which such an unequal power relationship exists, there is a possibility of domination or paternalism. The expert, who by virtue of his or her superior knowledge and the values of his or her professional identity "knows best," feels compelled to act on the patient's behalf but not at his or her behest ([41], p. 21; [87]). The attitude may develop (may even become part of the profession's natural attitude) that the more educated one is, the more responsibility one has to care and decide for the less competent person. The gap between the uninformed and the expert feeds paternalistic attitudes, and can produce an attitude of cynicism and resentment that works to devalue the therapeutic relationship.[16]

The fiduciary component of the professional accent rests on the premise that the physician is trustworthy. As previously discussed, trust is present if a person is vulnerable to another whose behavior is not under that person's control, and if the relation involves a situation in which the penalty suffered if the trust is abused leads the person to regret the action. Trust presupposes decision making in risky situations, where risk is attributable to the behavior of others or to the possibility that they will take advantage of circumstances for self-serving purposes. Such opportunistic behavior can mean stealing and lying, but also more subtle techniques such as withholding information to confuse or control.

Trust also presupposes that the action and, hence, the risks are avoidable. In the case of illness or accident, the person is thrown into a situation in which he or she may have little or no choice: "I have to trust this doctor, I have no choice." With no choice, we cannot invoke trust to explain our behavior. Whether or not a relationship is seen to be avoidable is often highly subjective, and presumably varies with the structure of institutions and their social power. The medical institution encourages patients to trust the individual doctor and doctors at large. It does not encourage second opinions or advertising. It accepts a "holistic" or "folk" approach to health care even less.

Finally, under the biomedical model's influence and the general acceptance of the scientific province, social classification shifts from stratification to functional differentiation. People no longer see themselves placed in a fixed social setting. They see themselves as having access to multiple functional sub-systems upon which they simultaneously depend. Essential structures and territorially bounded cultural entities are largely displaced by time-limited entities such as fashion and style. As we have seen, scientific activity encourages functional differentiation. Scientists also surprise the public with new discoveries and new theories as a matter of routine. These new conditions and relationships, of opportunity and dependence, of openness and lack of integration, have an effect on the structures of confidence and trust. Trust remains vital in interpersonal relations, but particularly in functional systems like economics, politics, or health care, trust is no longer taken for granted as a matter of course ([169], p. 104). Great effort has been spent developing procedures in these enterprises, but while procedural systems may create confidence, they do not elicit trust.

The promise of science and technology to narrow the gap between unlimited demand and scarce resources has been unfulfilled. It is difficult to accept the idea of progress in medical matters as an optional goal and not an unconditional commitment. If everything cannot be done for everyone, we must decide who can do what for whom. Decision-making, acting responsibly in the present, is necessary, but there is no knowledge beyond the moment by which we can guarantee the outcome in advance. Every decision is based on past knowledge, a perception of present needs, and is made with an eye to the future. Yet every decision remains inexact and its outcome ambiguous ([41], p. 116).[17] Thus, scientific medicine is a human activity no more absolute or relative than any other.

In sum, the breakdown of the trusting therapeutic relationship has led to an attitude of dis-ease between the physician and the public (represented by the patient). Not all patients distrust all doctors, not all doctors distrust and seek to control all patients. However, relations between people and the profession are in flux. The accent of reality that shapes the expectations of the participants in the health care relationship is characterized by this suspicion. More specifically, trust in this climate of suspicion has become a mechanism used by the profession to acknowledge public concerns while maintaining the profession's control of its work. Trust in this climate becomes an illusion with no substance.

It is not coincidental that as the scientific biomedical model has led to an increase in expertise and technology, there has been a corresponding development of public concern with health care (the development of bioethics as a discipline reflects this concern). For instance, in recent years medicine has been criticized for de-emphasizing the patient as person [90, 229]. While advances in medical knowledge and technology result in significant gains with regard to treatment, patients feel alienated from their physicians. Much effort has been made to make medicine more humanistic [246]. While fewer practicing physicians find the biomedical model satisfactory in caring for patients, much of the recent literature in bioethics continues to argue for physicians to develop a "human vision," as opposed to a "medical vision," to better care for the patient.[18] The medical profession has given attention to regaining the confidence and trust of the public in recent years by acknowledging the public's concerns about cost and control of health care, humanistic medicine, and the rights of patients. However, the biomedical model of medicine remains dominant and determines the foundation of the medical province of meaning.

The physician cannot escape a "medical" view of himself and his relationship with his patient due to the phenomenological boundaries (the horizon of meaning) of the medical profession. This view serves to characterize the meaning of medical "work" and expectations of trust for the doctor, even as it serves to separate the phenomenological reality of the physician from the patient's experience. While the two social roles are correlative, they are phenomenologically distinct in the way they conceptualize and "trust" their experience and each other. A phenomenological "shock" occurs when expectations are contradicted, when what was trusted or taken-for-granted is called into question. Distrust is a method used to restore a sense of control in the interaction. If the challenge to a horizon of meaning remains unresolved, this distrust becomes destructive of an understanding and acceptance of the other. In a self-protective posture, reform of the relationship becomes shallow and illusory.

NOTES

[1] The first meaning of trust is from ([8], p. 14). The second meaning of trust is what Talcott Parsons calls the "other-orientation" ([214], pp. 457–467).

[2] Talcott Parsons argues that it is in the physician's self-interest to act contrary to his own self-interest in an immediate situation since the patient's vulnerability makes a certain measure of trust necessary for the relationship to function ([214], p. 473).

[3] The "physician as such studies only the patient's interest, not his own" [224]. For Plato, physicians did not have a special duty of benevolence (love of humanity) towards patients. Such an understanding comes from the Stoics. Nonetheless, as physicians they necessarily had a special moral duty of beneficence – of doing good for patients (see [96]). Thus, it is not part of the concept of "physician" for doctors to have a love of humanity (although many do have a basic sympathy for the sick), but it is part of the physician's accent of meaning to have a duty to help the sick. "Service to clients," of course, is not a profession's only moral objective. Such an attitude does not always take precedence over other considerations (including other moral considerations such as the duty to the profession or a duty to the law or to society as a whole).

[4] Not all possible types are "present" before the physician. For instance, a doctor's own patients and associates, who are encountered face to face, are "consociates." Second, while not known by the doctor personally, "contemporaries" are other people who may become patients or members of the medical profession. Third, "predecessors" may be known to the doctor in the past as consociates, contemporaries, or historical figures (who have helped to shape the world in which he or she lives). Fourth, the physician's actions bear indirectly or directly on future consociates, contemporaries, and those who will be "successors" in the profession. Finally, there are those who are effective in the doctor's world although they are mythical or fictional characters whose actions and thoughts can influence his or her life, and the expectations of doctors on the part of others. For a discussion of consociates, contemporaries, predecessors, and successors, see ([293], p. 38). Fictional characters can affect the living, e.g., the T.V. shows "Marcus Welby" and "Dr. Kildare" revolved around the personal and professional lives of fictional physicians. The popularity of these characters helped shape the image and expectations that people have of doctors and of what medical care "should be," and, therefore, had an effect on the profession generally. Indeed, Drs. Welby and Kildare also may have served as role models for people who decided to become doctors.

[5] Medical history-taking is another example of the constraints imposed by individuals' biographically determined situations. Although we are consociates and contemporaries in time and place, in the health care relationship the patient's biographical situation is open to the doctor in a way that is not reciprocated.

[6] I do not mean to suggest that a doctor who appreciates that medical knowledge cannot be perfectly mastered, or who recognizes the limitations of medical knowledge, will succumb to uncertainty. However, the physician who seeks the certainty of complete knowledge is controlled by uncertainty.

[7] The young physician must come to trust himself or herself and his "knowledge" of his "work" and patients. The purpose of medical education and socialization into the "clinical" perspective is to foster within the medical student the profession's system of relevances (see [37], pp. 82–83).

[8] A similar interpretive procedure occurs in matters involving a fiducial relationship. A person may choose to trust another, interpreting actions according to a "normal form," until the epoché of distrust can no longer be maintained or reconciled in the face of contrary evidence. Once a situation is identified according to the normal forms provided by the medical province of meaning, due to the "reciprocity of perspectives" a physician assumes that when presented with similar data, another doctor will concur. The professional's opinion, if reached through normal forms, "should" be shared by any other who sees it. This

expectation often is extended to non-professionals as well. If a patient asks for information, doctors often provide a set of explanations or directions grounded in the understanding (or clinical perspective) defined by their province of meaning. The assumption is made that the patient shares with the doctor "constant objects" (or a mutual understanding).

[9] Such commitment can also lead to paternalistic attitudes on the part of physicians. After all, to the doctor who is convinced of his or her own trustworthiness, "enlightened" patients will acknowledge the superior expertise of the physician and follow his orders. For the unenlightened patient, paternalism is the only viable alternative; see the reference to Benjamin Rush, in ([149], p. 16).

[10] The natural attitude of everyday reality acts to create a social cohesion without which we would be forced to question or doubt every aspect of our lives. The constraint is relevant not only for us in deciding how far we need to trust others, but also for others to decide how far they can trust us. It is important to trust others, but it may be equally important to be trusted (see [92], p. 221).

[11] See Thomas Kuhn on the relationship between normal science and crisis science [160].

[12] Technology means more than the development and utilization of machinery. It designates an aspect of culture that can shape the institutions and the morality of a society. For instance, changes in medical technology, such as saline abortion or vacuum extraction of the fetus, effected changes in the law regarding abortion. These technological changes coincided with changes in public morality that made abortion acceptable for many (see [37], pp. 69–70). However, more recent advances in neonatal technology effect the moral understanding of "viability" and may create a review of the legal basis for abortion (see [41], p. 100). Generally, every society is influenced by technology. We have become a more self-consciously scientific and technological society during this century. As a result, we are guided by a worldview governed by a techno-economic value system (see [279]). There is an corresponding effect on *embodiment*, on the ways in which we view the everyday world and our bodies as part of that world.

[13] Ivan Illich writes of the medicalization of society in arguing that physicians have extended their area of "expertise" beyond its appropriate limits into other areas of social life [141]. However, the phenomenon of specialization acts to narrow a person's focus to a part of the whole even as the specialist assumes that he or she is an "expert."

[14] Combined with the increased range of choices in health care, one result of the emphasis on informed consent, patients' rights, and autonomy is to place the burden for choice more fully upon the patient or public.

[15] The number of subspecialties is booming ([49], p. 12). As the fields of knowledge increase, the expert becomes locked into a particular specialty. His or her horizons become shaped by the limits of his or her expertise. This movement to specialization within medicine produces physicians who are "typical" of their specialty. Interested in a particular facet or problem of health care, these doctors typically concentrate on a structural or functional part of the whole patient. However, "specialization per se does not exclude the art of medicine as an ingredient in its practice; but essential to the fruition of the art is an appreciation and understanding of the whole patient" ([39], p. 86).

[16] Since it is the doctor's responsibility to act in the patient's best interest and if patients accept and want this, it may be argued that this gap is inevitable and works to benefit the relationship. Dostoevesky makes the point that people are afraid of "true freedom" and seek someone to make decisions for them (see [60], pp. 227–245). Others argue that the rise of patient autonomy is a reaction against paternalistic medicine's monopolistic tendency (see, for example, [18], p. 63). Unfortunately, the tendency in the face of unclear lines of authority and ambiguous possibilities is a retreat into ideology. The true anguish of bioethical issues often is obscured by the movement to the clear and easy answers offered by ideological "camps." In most decisions, someone stands to gain in some way while someone stands to lose – this ambiguity haunts all decision-making (a choice of one option eliminates the choice of another option since we cannot have everything we want). Those

immeshed in ideological certainty refuse to acknowledge this ambiguity (see [98]). There is also the possibility that "experts" will disagree (see [177]).

[17] It must be acknowledged that science can be a powerful agent for questioning the taken-for-granted assumptions that nurture it ([106], p. 21). The natural science of the 19th century grew out of cultural assumptions that nurtured Darwin and Freud, yet both of these "scientists" presented revolutionary challenges to their culture.

[18] Oliver Sachs discusses "human vision" as directed at the person who is ill and "medical vision" as directed at the clinical data of illness ([239]; and [10], pp. 606–611).

CHAPTER 5

TRUST-AS-CONTROL IN A THEOLOGICAL PERSPECTIVE

A key tenet of the medical profession is that physicians work for the benefit of their patients. However, by doing so they also work for their own benefit, whether this benefit is financial, altruistic, or the self-satisfaction of a job well done. Growing out of the mutual dependency that characterizes human relations, because self-interest is always an ingredient in the physician's work, physicians carry the seeds of alienation and estrangement from others into their work. If unacknowledged and unappreciated, and not balanced by a realistic awareness of the dynamics of power that characterize the medical relationship, self-interest can foster the conditions that give rise to an inauthentic "giving and receiving" of trust between physician and patient.[1] Such giving breeds suspicion, resentment, and shame within the receiver of care, even as it demeans the caregiver. It is produced by the desire to escape dependency, yet such a desire only serves to perpetuate that condition.

In the shock accompanying illness and the recognition of dependency, the loss of the epoché of well-being threatens one's self-identity and control, and disrupts previously taken-for-granted relationships. The phenomenological *breaching* or interruption of the natural attitude that illness provokes suddenly makes one aware of experiencing "himself" or "herself" as dependent upon others, that he or she is both subject and object in the world. When ill a person is aware of the fragility and vulnerability of the *self.* Such awareness naturally raises a desire to minimize vulnerability by restoring a sense of self-control.

In the clinical encounter, however, typically the physicians' and patients' self-interests reach some sort of accommodation. A part of the physician's task in treating patients is to bring them to alter their perceptions of their life-world, to accept the doctor's understanding and explanation of the patient's symptoms. This alteration in perception allows the patient to place the experience into a meaningful perspective based on a common frame of reference (one presented by the physician). While all social relationships involve an exchange of power and control, even in a mutually balanced interaction, from time to time one party has a controlling influence in the "give and take" of relationship. When either party becomes a controlling presence, the common universe between them is fragmented.[2]

Unfortunately, the habits of mind that biomedicine fosters in many physicians too easily are disconnected from patients' experience of illness. In the medical province, armed with medical knowledge and socialized into a world removed from everyday life, the physician may slip into perceiving the patient as a *body* over and against the doctor's self.[3] For some doctors and patients, the claims of expert knowledge and efficiency may compensate for a distant, "professional" relationship, especially if people have learned to expect that model in their clinical encounters. However, Leon Eisenberg claims that "present-day disenchantment with physicians, at a time when they can do more than ever in history to halt and repair the ravages of serious illness, probably reflects the perception by people that they are not being cared for" ([66], pp. 235–246). If a patient finds that he or she is not "cared for," the patient's sense of vulnerability is heightened, and his or her desire for (self)control strengthens. As patients have increasing difficulty determining the relevancy of biomedicine's approach to their experience of illness, they take up distrust as a means of controlling their situation; to ensure their authority, physicians use trust in their expertise, bolstered by "facts" and experience, to control their work. Such trust-as-control threatens the give and take necessary for a cooperative medical encounter.

In the broad, popular Christian tradition, the image of the transcendent, creator God provides a central focus for a sense of meaning and reality.[4] It is assumed that God exists or has reality distinct from and independent of human nature and our way of thinking about God. In the postmodern age, however, the relevancy of God to human life is an open question for many people. Once the notion of God-as-center is bracketed, the human community is thrown upon its own resources for identity and coherence. The effort to replace theology with anthropology has given rise to humanistic atheism. In the postmodern age, such atheism also has proven unable to provide people with a sense of unity and community that transcends their perception of finitude. The result is a culture of criticism in which nothing seems sacred. To many it seems that we are standing on the abyss of nihilism [110, 270].

In a culture flavored by uncertainty, uncomfortably aware of the impermanence of values and suspicious of authority, the person confronts without mediation the riskiness of relationship, the perception of the precariousness of individual and social identity and security. In an attempt to guarantee self-identity, one is tempted to restore a sense of self-control and autonomy by insulating oneself from risk and uncertainty. However, while this movement of self-preservation is natural, it represents the desire to eliminate what is inescapably present – that the human need for relationship is a risky business.

In the climate of suspicion created by the desire to escape risk by denying neediness, both parties in the health care relationship react from within a particular horizon of meaning or accent of reality that is increasingly characterized by self-protectiveness and defensiveness. The danger of such a climate to social relationship lies in a tendency towards skepticism or the use of distrust as a means for social control of others. The patient who reacts to the need for additional medical tests by announcing "All you doctors are alike. All you want is more money" may be using distrust to assert a measure of control over the situation. Such an attitude seeks to manipulate the doctor into some confirmation that he or she is trustworthy – that he is not like "other" doctors. This manipulation can produce feelings of anger, resentment, and shame in the physician. In this chapter, a theological analysis of the use of trust-as-control by physicians to dominate others in the clinical encounter highlights its debilitating nature and inevitable failure as the basis for the medical relationship.

THE POSTMODERN BREAKDOWN OF FAITH IN A TRANSCENDENT GOD

The Enlightenment represents a cultural-historical development of this-worldly values, methods of inquiry, critiques of traditional authorities, and criteria for truth that claimed to cut through the conflicts caused by the ambiguities of historical religion. As a reaction against authoritarian religion, the *philosophes* shared an optimistic belief that humans themselves are able to provide for the needs of the human spirit in an atmosphere of freedom and reason [94]. Moreover, this spirit brought changes in ethical and social theory, characterized by an emphasis on humanitarianism, individual rights, and an appreciation for the critical method of inquiry. It favored an appreciation of the role in structuring cultural meaning and values played by individuals in their daily social interaction.

The application of the critical method to the study of human life resulted in the development of the social sciences. Social scientists maintain that reality is socially constructed, a process played out within the world-field of objects and persons external to the individual's consciousness, affecting in turn his or her way of thinking and acting in the world ([244], pp. 3–19). For sociologists and other social scientists, this interaction is a given aspect within existence. Therefore, we do not derive meaningful patterns of life simply through conceptual thinking. Instead, we become familiar with the world by living in it, encountering and struggling with other persons and objects. It is through this active participation in the world that conceptual thinking and patterns of living develop. These patterns then are internalized and come to affect our activities within the world. Just as social scientists conclude that there is no universal self apart from the body, so the embodied self does not live in isolation but within an objective world and an interpersonal matrix. In a sense, everything we know of the world is received from our social milieu. All is presented to us through the social makeup of the everyday life.

In sociological theory, morality, as a sense of value and meaningful existence, and ethics, as a means for evaluating behavior within morality, develop over time within the reciprocal relationships of society's members. Ethical thinking helps shape and maintain the ties that bind individuals to each other and to the community. An individual or group's behavior and choices regarding appropriate behavior arise from the patterns of life deemed meaningful by a community. As long as these actions can be assigned meaning, and, therefore, be made acceptable within the limits of such a "reality," some sense of harmony is maintained. No appeal to an authority of normative behavior outside of the system is considered.

Applying this idea of a socially constructed reality to religion, religious concepts are externalized or projected, given a facticity, and internalized as part of the ecclesiastical tradition of the religion and as part of the ongoing history of a society. These concepts come to act upon a people by establishing moral and ethical norms and belief systems. In a pluralistic culture in which religions must share a place with other provinces of meaning, religious concepts, beliefs, and values are given meaning in the world of daily life.

In the post-Enlightenment period, the theistic God is conceptualized as omnipotent, omniscient, and omnipresent. Such a God is the Unmoved Mover who is the origin of all motion and the source of all rest. The God who is alone God is "an identity which is only itself" ([2], p. 19). Therefore, God is always more unlike than like our best utterances of him. As a result, God is intrinsically, unavoidably, beyond us ([38], pp. 140, 142). The deity is fully and completely self-enclosed and self-present. This ultimate subjectivity of God does not establish a continuum between divinity and humanity. As the full realization and original ground of selfhood, God is wholly *other*, is absolute *alterity* [(267], p. 23].[5]

The conceptualization of God as *other* reflects a polarity that is the basis for the theistic model carried within the Western ontotheological tradition. Though consistently monotheistic, Christian theology operates with binary terms. Whether speaking of God and world, infinite and finite, life and death, giving and receiving, or trust and distrust, these opposites are not regarded as equivalent. Invariably one term is given an ontologically privileged position through the reduction of its relative (e.g., "It is better to give than to receive"). The resulting economy of privilege sustains an asymmetrical hierarchy in which one term rules the other throughout the theological and axiological domains ([267], pp. 8–9).

This dyadic formulation, in which each term has ontologic value of a primary and secondary nature, extends to the domain of health care. The conception of relationship places the terms "physician and patient" and "health and illness" in a seemingly exclusive and evident opposition. The first term in each pair obtains a primary position in the ontological economy of privilege. It is this economy of relationship, both in its social and conceptual ramifications, that has led to the cultural climate of suspicion that undergirds social interactions generally, and the doctor-patient relationship in this discussion.

The spirit of the secular age is marked by an emphasis on the world as shaped and structured by human reason and will. However, in the midst of the gradual secularization of Western society, it also is clear that the way in which we view the world and organize a meaningful existence depends upon the model of meaning and structure of reality to which we subscribe (in which we have faith). There are many models available in the world, each claiming to represent reality most adequately by defining experience in the most meaningful way. Whether we examine a particular religious, scientific, political, psychological, or economic model or paradigm, each represents a "reality" according to its own terms and truth claims ([156], pp. 7–8; [160, 264]). Any one can provide norms for behavior that take on a particular meaning and value consistent with the parameters of the model.

While a model may be consistently meaningful within itself, thereby offering some means of identifying "norms" for behavior, it cannot extend its reality claims beyond its own parameters.[6] Existing in a world made up of multiple provinces of meaning, a person may be able to ignore these competing claims for a time, but finally, inasmuch as these provinces are finite, anomalies appear to challenge the province's accent of meaning. As a result, ". . . the spirit of the typical 'modern' man is relativistic and skeptical. . . . What do arguments prove, if an opposite theory can be based equally logically upon an impregnable principle?" ([27], p. 17). If the person is to possess a grounds of authority for lived experience in a world in which various models compete with one another, some principle must be found that can transcend the truth-claims of competing models and offer a means for resolving relativistic interpretations of "good," even "normal" behavior. The concern with moral relativism, or the inability to determine a ranking of moral principles and theories, reflects the difficulty of establishing such a principle.

The problem goes deeper than recognizing a plurality of models or paradigms. Phenomenologically, habits of mind constitute experience – construct it – in ways that are not quite captured by concepts like models or paradigms. Because the notion of paradigm offers a convenient way to talk about how we orient ourselves in relation to the world, the term has become central in descriptions of professional practice. Paradigms, however, result from habits of mind – of observation, of selection, and of interpretation – that are the bedrock of a person's "approach to the world." Medical paradigms and models represent ways to organize a profession's culture, but these are themselves constituted by spontaneous and taken-for-granted habits of mind ([156], p. 8); we "live" them before we "know" them.

Prior experience and the meanings attributed to it affect present experience, often in subtle as well as determinate ways. Through our habits of mind, meanings are present but not visible in a situation; they are of "second nature" to us. This second nature, its naturalness and spontaneity, accounts for our ability to "make sense" of reality. When habits of mind are called into question, a person's attitudes towards everyday life, his or her understanding of the past and present, and his expectations for the future, suddenly are adrift. It is this spirit of deconstuction characterizing the postmodern age that challenges the privileged priorities in all relationships that formerly served as a basis for taken-for-granted truths. If there are no foundational truths to serve as the basis of our relations with one another, we would seem to be left merely with relationships of mutual self-defense and self-interest.

There is in the postmodern world presently a crisis of trust. The theistic habit of mind that determines Christian observations, selections, and interpretations of reality is seen as incommensurate by other habits of mind in a pluralistic culture. However, the "unambiguous" status of science and its praxis was supposed to eliminate metaphysical and epistemological confusions and ambiguities attributed to religious thought. Historical, sociological, psychological, and phenomenological critiques of religious authority have produced religious, ethical, and cultural pluralism, and instead of eliminating have created new metaphysical and epistemological interpretations. As a result, Western culture is passing through the culture of criticism into a postmodern period in which conceptions of science and religion both are undergoing criticism.

The effort to establish certainty through the Cartesian method of doubt and the scientific search for an objective basis for confident knowledge as a basis for action in the world leads to "superficial" knowledge at the cost of a depth of meaning (see [295], pp. 126–132). It was believed that the proper formulation of method would uncover the limits of natural law and reveal the inner workings of creation. Such a viewpoint presumes that the world of facts, and often the world of values, has a unique pattern of rational coherence ([70], p. 19). The emphasis on knowing thoroughly, on seeing all sides, represents the desire for fully objective knowledge (knowledge of the *ding an sich*, the thing itself) – a vestige of Descartes' wish for certainty and the Enlightenment confidence in universal reason. However, the scientific method, which has been employed in this drive for a universally common perspective, is unable to deliver the goods. With the movement from skepticism to nihilism, the question is no longer what can we know, but can we know, or depend upon, anything at all?

Thus, the late 20th century is a time of growing distrust, not only in religious belief, but in the belief structure of science and humanistic culture [95, 109]. For instance, in the secularized natural attitude, science is believed to need no external moral guidelines. The guidelines of science, however, have proven inadequate to secure deeper human values. While science and technology have become indispensable to our lives, we are experiencing the shock that occurs when confidence in a preceding ethos, such as the ideology of progress, is shaken [208]. A yearning for reintegration of meaning and value to set the world right and reduce anxieties leads to the recognition that the scientific accent of meaning is only one aspect of the world of everyday life. This perception reflects a criticism or epoché applied to secular culture's accent of meaning. It opens the way for a return to a theological analysis of the nature of dependency and the problem of an economy of domination in relationships. Issues of domination and exploitation arise in human life as we attempt to equalize the unequal by erasing the difference of that which is *other* – including the other-God ([267], p. 28).

THEOLOGICAL ROOTS OF TRUST-AS-CONTROL

One theological tradition within early Christianity claimed that Jesus suffered in his "human" aspect, but that this spirit was distinct from his "divine nature."

According to Arius, the suffering Jesus was not God, not even a symbol of the nature of God, since the divine nature excluded all suffering. He argued that God is a selfless divinity, above all needs. The only response to the perfection and absolute will of this God was unconditional obedience in a life of faith. This claim was based on the assumption that God neither needed nor sought the mutual good of fellowship with humanity, yet choose to relate by the gift of grace in Jesus. God retained the privilege and the power of the "free" gift of relationship. The ultimate inspiration for the idealization of selfless giving was this vision of the divine as pure self-giving devoid of all self-concern ([182]. In the Arian tradition, the *imitatio Dei* becomes one of abnegation and the diminution of mutuality and reciprocity between the human and the divine.

Classical theism denied the relativity and temporality of God. At the same time, the tradition held that Jesus was both suffering man and God. The Patripassionists drew the conclusion that God must suffer. The orthodox reply to this position was that only the manhood in Jesus included suffering, not the divine nature. The two natures were really distinct, even though of one person. According to orthodoxy, the suffering Jesus was not God ([116], pp. 152–153).[7]

Although soundly denounced by Athanasius and his followers, the Arian view of God as wholly self-contained lingers as a counterforce in Christian theology that has surfaced in the modern period ([182], pp. 34–52). The claim that the transcendent God of Christianity is experienced as domineering rather than life-giving to human individuals and community is disturbing. Generally, the classical Christian theological tradition perpetuates the priority of the divine center over the individual human, and relations between individuals in community are expected to model that relationship. Knowledge of self is mediated by knowledge of God, and humans are expected to imitate God-as-Christ in their dealings with others.

However, in the postmodern world, attempts to bolster the hierarchical relation of opposites that sustains the theological economy of privilege fail to fully satisfy [238].[8] As long as the assumption of a hierarchically based *imitatio* is maintained, people are left with shame at their stubborn selfishness and inability to match the idealization of selfless giving. In the attempt to bracket discomfort and doubt, people are driven to seek self-sufficiency or self-annihilation.

The depths and tenacity in the postmodern world of the hermeneutics of suspicion must be appreciated. The denial of God as central to human life and the elevation of humankind to centrality throws individuals into confrontation with other selves, and finally with the otherness within themselves. At the moment when the phenomenological epoché is bracketed, the self confronts the other within and without, and self-identity is disrupted. God and self become master and slave, engaged in a life-and-death struggle that is inspired by the shock that grows out of a direct encounter of the other as *other*.

The Christian theological tradition maintains that humankind is created in the image of God, and, therefore, is bound by its relationship with God and each other. If God and mankind image each other, as in a mirror, the self sees itself reflected in the other. However, the encounter with the other as *other* leads to the encounter with the *self* as other in the reciprocity of perspectives. Facing the all-powerful master, the self realizes that it has lost itself, for it only finds itself as an-*other* being ([267], p. 23). By forcing the self outside of itself, God discloses the subject's estrangement and self-alienation. In the Augustinian tradition, this disruption or confrontation is the necessary first step to reconciliation with God in faith. However, in a climate of suspicion, awareness of their dependency drives people to seek self-possession through the domination of others.

Dependency and Domination

Social existence involves the gift of life, of meaning, and of place in the world. A person born into an ongoing tradition is given an identity and a means for making sense of reality. It is crucial to note that giving represents only one-half of a person's relationship to the everyday world. The giving one returns to the world completes identity. It is through this exchange that people participate in "reality." The natural attitude of the world of daily life into which they are born represents (or gives) a sense of reality to them, and they in turn re-present (or give back) themselves to reality. Therefore, persons participate in reality by giving and receiving through combinations of relationships ([289], p. 255).

Ordinarily, these combinations occur in a taken-for-granted fashion as the person lives in the world, assuming that the conditions of his or her life are as they should be. However, when the natural attitude is disrupted, the person becomes aware that his or her life could be other than it is. The person becomes aware that he or she is incomplete, dependent upon others for the conditions that make up his or her "life."

In any relationship within the everyday world, giving and receiving are involved. Furthermore, receiving depends upon the nature of the giving. Ideally, receiving from the world (and from others who are a part of it) should not become a means for alienation and anxiety. Unfortunately, in the world of daily life in which people struggle, compete, choose, demand, punish, and reward, rarely are gifts exchanged freely. As people begin to suspect the depths of their dependency, they become possessive of their identities, suspicious of the other, and threatened by that which is outside of themselves. The presence of *otherness* creates anxiety, and seems to produce a defensive withdrawal from the other in order to protect and preserve the self ([289], pp. 228–229).

Dependency causes people to guard themselves jealously. For example, the power of medicine to sustain life is often a source of fear rather than solace to people with chronic or terminal conditions, experiencing great pain and suffering. They become suspicious of others' motives and resist receiving the gift of relationship with gratitude because they fear becoming obligated (indebted) to the giver ([33], pp. 1–34). Consequently, people cannot receive easily and cannot give easily in return. Gifts become commodities which have a price, jealousy retained or carefully exchanged to maximum advantage. In this attitude, neediness and dependency are perceived as weaknesses that people are afraid to reveal, yet these conditions lie at the very root of individual and social existence.[9]

In this condition of dependency, people tend to objectify the other, reducing his or her gifts, and therefore his or her self, to the status of "objects." Domination and exploitation become motivating forces in human interaction as people attempt to mask their dependency from others and from the self. Through a person's fear at being vulnerable, he or she renders himself or herself unable to accept life in its fullness (including illness, suffering, and death) as a gift, freely given and freely received. Consequently, he or she cannot live his or her life with gratitude and feels threatened by the need for relationship.

The natural attitude within which people ordinarily encounter their world carries with it a sense of permanency, of *real estate*. Unlike possessions, which are removable, able to be made absent when the tax assessor arrives, *real* estate is defined as that property which is fixed in place. Property, therefore, becomes what is fully present. The economy of domination is tied to this logic of property in Western metaphysics. Confronted by a god who is absolutely beyond their control, fearful of that which is beyond their control, people seek to imitate aseity and thereby gain control over their creatureliness. However, this control becomes self-directed as the person seeks to acquire and retain some "fixed" meaning in life. A drive towards domination of others is based on the principle of ownership, which, in turn, is based on the acquisitive logic of need ([267], pp. 27–29).[10] Too easily, this drive seizes upon one's body and sense of self as a possession to be protected. One's "well-being" similarly becomes well-guarded.

A sense of physical well-being and personal integrity allows a person to "work" in the world in existential and ontological peace of mind. Illness is an interruption, a bracketing of a person's sense of embodied wholeness.[11] Illness names that experience in which our everyday, embodied capacities fail us. It obstructs our ordinary access to the world and presents the body as a signifier for the way in which we are limited and dependent beings:

> Illness erodes the image we have constructed over the years, often painfully, of ourselves and our world. That image is our personal effort to harmonize our deficiencies and our strong points. It is our personal definition of our situation in respect to others and the world, our unique relationship to work, play, or salvation, laboriously fabricated and balanced against the changing exigencies of life. Illness threatens this carefully wrought self-image ([222], pp. 158–159).

Pellegrino claims that people fashion existence out of the circumstances in which they find themselves. Each person confronts the realities of his or her unique situation and humanizes it, attaining an equilibrium with it. "In short, the reabsorption of circumstance is the concrete destiny of man" ([222], p. 159). Illness, however, moves us toward the absorption of a person by circumstances. Illness fragments our constructed world by bracketing the natural attitude that enables us to organize and "live with" circumstances and exigencies of human existence. It is this fractured condition that the medical relationship works to "heal" by a mutual effort with the patient to find a new balance, to restore a balance or an integration of the changed circumstances that enables the person to recover some sense of his or her personal project ([222], p. 159).

However, in the movement toward the absorption of the person by changing and threatening circumstances of illness, a person's self-control is disrupted. The person is made vulnerable in relations with others. The power differential between the physician and the patient is, in effect, an economy of domination that results in a focus on utility and consumerism in the health care relationship. According to the principle of ownership, the doctor "owns" or possesses his or her knowledge and skills, the patient "owns" or possesses his or her body; each possesses certain proprietary rights. Health care becomes a bartering battleground in which goods and services are exchanged in an process of giving and receiving cure and care. In recent years, there have been controversies regarding a person's right to buy and sell human organs (does one possess ownership of his or her body and have the right to sell pieces of it), surrogate mothering, and physician investment in laboratories and testing facilities (after all, their knowledge is their possession and can be utilized for additional profit) ([247]; [6], pp. A1, A4). These episodes are upsetting, but their notoriety reveals their valency; they represent efforts by participants to acquire and accumulate power to minimize what they give and maximize what they receive. However, dependency and neediness remain and cannot be eliminated. These two conditions may be ameliorated, but they cannot be controlled.

The modern struggle for autonomous existence free from the domineering presence of a transcendent god finds expression in a drive for "health" and "life" as commodities for self-possession. The demand that there be equality in all respects between the physician and the patient, or that everything be done for a patient, is a reflection of this struggle. However, the effort to possess the other represents an indirect attempt to possess one's own self. The attempt is doomed because it carries within it the seeds of its own destruction. The effort to possess total and complete presence, without loss, without need, results in death. Although we strive to possess ourselves as property, it is not the presence of self that we can control. It is the recognition of an absence of possession and control that gives complete meaning to the self.

Shame and Resentment

In the absence of full and final self-control, with the recognition of dependency as incompleteness, a person often feels guilt or shame in relations with others, somehow finally inadequate. The two differ in important ways. People who feel guilty invoke the concept of right and wrong, and expect their victims to be resentful of the action producing the guilt. Although guilt results from injuring or violating the good in another through the denial of one's own need, at least to feel guilty is to be willing to live in relationship. The person is willing to receive from another, although he or she might violate the gift and feel guilty. "Guilt" can be relieved by reparation and forgiveness that permits reconciliation ([42], pp. 583–584). However, shame is a feeling that a person has when experiencing damage to his or her self-esteem ([231], pp. 442–446; [147], pp. 40ff). People who are ashamed appeal to an ideal of which they have fallen short. In a discussion of the use of trust-as-control, shame is the more troubling concept.

Shame results when alienation has reached the point that the person feels deficient in his or her *self.* Every encounter threatens to remind the person of this deficiency and reveal it to others. To mask this feeling, people may react in a variety of ways. First, those who are ashamed may strive to become selflessly involved or preoccupied with the world in order to deny their violated sense of self. Second, people strive to destroy the goodness that forces them to confront the inadequacies within themselves by turning outward to dominate the world around them.[12]

If either of these approaches is taken, people engage in a dynamic of giving and receiving twisted into a self-centered giving and shameful receiving. If the giving that occurs in relationship is of a type to evoke shame, the interaction will be received with resentment. Shame fires the desire to gain control over the oppressor and become masters ourselves. It drives us towards absolute autonomy and independent selfhood as the means to relieve our estrangement and humiliation at being deficient and needy.[13]

The possibility of shame and resentment in the doctor-patient relationship is a seldom discussed topic. Any illness can be shame-producing for the patient and the physician [163]. Illness may bring on feelings of helplessness, create changes in one's mental status, and may produce physiological symptoms that are distasteful or offensive to the patient and others. The person is aware that he or she is "set aside" from the "healthy" condition that now belongs to others. These perceptions can lead to feelings of shame and humiliation. The physician, too, may feel ashamed and humiliated if there is a diagnosis failure, a poorly performed procedure, if the patient or family is disrespectful, or if the physician identifies psychologically with the patient's condition (i.e., a rapidly deteriorating dementia and the patient's resulting loss of capacities). In the condition of shame and humiliation, the person will acquiesce to or resent the one whose presence evokes the feeling [195].

If in relation to illness both physician and patient become aware of *otherness*, either in awareness of their own embodied being or in awareness of their dependency on others, each struggles to eliminate this awareness in two seemingly contradictory ways. On the one hand, each one is attracted to the other and seeks identification and/or incorporation with him or her. For instance, someone who is ill may identify with the physician and follow his or her orders with slavish devotion. Also, physicians can identify so closely with their patients that their work becomes centered around the person's condition. On the other hand, the person is repulsed by what the other represents and attempts to negate or exclude the difference ([267], p. 29). The patient may reject the help which the physician represents in an attempt to maintain a sense of control. The physician who sees too much of himself or herself in the patient may adopt a distant attitude. In either effort, the other is converted into a means for each other's self-control.

According to the profession's tradition of ethics, which reflects the profession's accent of meaning, a physician's commitment to fiduciary service is depicted as selfless service or giving for the sake of the patient. This commitment can become distorted under the conditions of shame or resentment. The other person in the relationship, the patient, is expected to respond gratefully by receiving the physician's gift of service (of time and energy, for example). The patient should trust the doctor since the doctor gives of himself or herself for the benefit of the patient. Thus, compliance and gratitude become the "payment" for the physician's gift. Of course, the physician controls this economy of exchange.

The physician may not be conscious of this kind of control. In the effort to dominate the uncertainties and the involvement with otherness that medical care unavoidably represents, the physician seeks a means for self-expression that at the same time shields him or her from the shock of confrontation with one's own neediness. An attitude of selfless giving leads to the "conceit of philanthropy" in which one gives to make his or her superiority apparent ([174], pp. 29–38). Such philanthropic giving also serves to make the giver feel good in his or her superiority. Giving becomes a means to deny both neediness and the shame that accompanies the denial of personal dependency. Also, a physician's anger at patients can suggest the physician's shame at not knowing, not doing more, indicating a feeling of inadequacy or self-deficiency.

Under these conditions the exchange of relationship between doctor and patient becomes a struggle for mastery that joins affirmation with negation. The self affirms itself by negating the other, in the extreme by reducing the other to negative terms (e.g., "gomers"). Consequently, the subject embodies a form of negation in which identity attempts to secure itself by excluding difference. This movement reflects allegiance to the noncontradictory logic of identity by marking the boundary of identification in an absolute manner. The subject tries to master alterity by negating the other and enclosing the self within the secure "solitude of solidity and self-identity" ([267], p. 24; [57], p. 91). However, the presence of the other threatens to introduce an awareness of a void, a need, in the subject. What develops in the relationship between physician and patient is a hierarchical relation of opposites in which giving and receiving becomes a battle for control; a battle to fill the void of incompleteness, of which the perception of otherness makes one aware.

According to Nietzsche, whether this dynamic of giving and receiving occurs in the divine-human relations or the social relationship, it evokes resentment. The effort to secure a self-contained self-identity by controlling relations with the other is a defensive one that grows from the underlying condition of dependency. In the hierarchical relation of opposites, this effort seeks a triumphant affirmation of the self but produces instead a reaction that negates what is outside, or other than, what is different, what exposes the incompleteness of the self ([205], pp. 346, 579). Rebellion against God reveals a deep resentment that sustains the disparity of power and control. Although the human is estranged from God, this estrangement is a form of relationship. On the psychological level, resentment is an ambivalent sentiment, combining affirmation and negation, acceptance and denial. It is never simply hostile for it carries within itself latent admiration of, and attraction for, the other against whom one, nonetheless, reacts negatively. The person's admiration is both hidden and revealed in the hostility.

Thus, resentment harbors envy, the counterpart of shame. This envy finds expression in a drive to transform the shame of being found inferior and being made to feel powerless into selflessness and service. The source of a patient's discontent is never simply the condition of bondage to the physician's power and authority. Although the patient is made aware of and may be resentful of the subservience of the "sick role," he or she yearns to possess the mastery of knowledge and position, the physician's power and authority (see [207], p. 36). The patient resents and envies the "health" of the physician who goes home from the hospital at night, and whose "presence" in the patient's vulnerable situation commands obedience. In our desire for health and wholeness (the desire to preserve the "I am"), the physician is invested with power, privilege, and authority to which the patient is subordinated. However, in a sense the physician is only a token of what the patient lacks (charts, white coat, technical armamenture, process and procedure – all indicate the absence of health, the lack of wholeness). For a patient the physician represents the *sign* of health, a link with the remembered, original condition prior to the disruption caused by the patient's illness. However, as Umberto Eco argues, such signs (e.g., the Hippocratic Oath, the CAT scan, or a white lab coat) deceive by claiming to correspond with the promise of health ([65], pp. 3–11).

The physician resents the patient who in illness represents a threat to his or her power and authority – the patient may not recover. At the same time, the physician also can envy the patient's vulnerability: Someone is "taking care" of the patient. Although the patient has much at stake, he or she does not have to confront the physician's struggles with uncertainty, with the anxiety of the faulty diagnosis or procedure, or with the shame and humiliation that can accompany the "loss" of a patient.

According to the professional ideal, buttressed by the public's expectations of self-sacrifice and altruistic service, the physician participates in the "religion of self-annihilation" by acting in the person's best interest.[14] Really, the physician's exercise of expert knowledge and authority simultaneously manifests his or her own strength and discloses the weakness that defines the patient. Often the doctor who works to withhold or overcome feelings of anxiety and emptiness by serving the needs of others is hiding feelings of resentment, envy, and shame. In the objectification of the other through self-centered giving and receiving of relationship, this effort at domination leads to a true self-annihilation.

The dynamic of shame and resentment fostered by inauthentic giving and receiving has another dimension. Demands for patient autonomy and reciprocity represent not only an attempt to share the professional's role as decision maker; they also represent an attempt to usurp his or her power. As a professional, the doctor naturally wants to retain control of his or her work. As we have seen, such control is what provides the doctor with a sense of meaning and identity. By achieving a measure of equality in the "work" of health care, the patient hopes to be able to overcome the lack, close the gap, or fill the need which disrupts and unsettles the sense of self.[15] Therefore, the struggle for patient rights and the movement towards the absolutization of autonomy in bioethics may represent the attempt to fill a lack of wholeness, to "possess" health by denying or negating the finitude and dependency of which illness makes people aware. However, since finitude and dependency are always present, the struggle to fulfill the void and negate the need fails. It becomes a recurrent struggle that forms the ground for a climate of suspicion and erodes the fabric of trust between patient and physician.

As consequence, the individual's struggle for wholeness, as self-possession, presently marks the relation between physician and patient. Both parties are dependent upon each other, and both react to this neediness by developing strategies for self-preservation through self-completion. The physician seeks the knowledge to overcome illness, thereby restoring wholeness to the other (and thus, to himself or herself). The patient seeks to control neediness by cooperation, denial, or aggressively demanding his or her rights. However, either person's strategy is only an effort to negate the correlate in the pair of opposites. The doctor's authority and the patient's demands, for instance, are reflexively derived from a negativity; that is, from our common fear of death. It is hard to focus inwardly on one's own condition of dependency. In the effort to overcome or deny the neediness that dependency creates, the easy target is the other in the relation. Trust in relationship is undermined in the drive for control.

THE THEOLOGICAL INVERSION

Phenomenologically, *world* represents that concept without which we cannot do, for it gives to experience the fundamental order, unity, and meaning apart from which it would not be coherent enough to be experienced at all. Theologically, the concept of God as the source and author of the world provides the first reference point of meaning and structure. The claim that "God created the world" means that the world gains its being and its fundamental forms of order from a source outside of or beyond itself. Thus, theologically, the deepest roots of the order within which life must be lived are not immanent in the world itself, but have been given to it from without ([267], p. 27). God becomes radically *other* as humans become created beings, wholly dependent upon the other for life, order, and meaning in existence.

Ironically, the struggle in which the subject attempts to assert itself by negating the other and securing identity by excluding difference inverts itself. It becomes an act of identification with and incorporation of the other. This movement is a reflection of Feuerbach's contribution to anthropological atheism ([151], p. 28). His inversion of the divine-human relationship can be applied to the one between physician and patient. The physician is the person to whom is assigned primary responsibility for defining, controlling, and caring for illnesses and those who suffer from them. Over time, this function shapes in an increasingly complex way the definitions and social roles that are associated with health care. Through the process of externalization, the public gives up powers to the medical profession in return for care and beneficence. This "gift" to the profession becomes objectified as health care and is returned to the public in the profession's fiduciary commitment. In the drive for patient power that has characterized health care in recent decades, the "inverted relationship" is inverted to gain what the patient desires: The attributes of the physician are transferred to the patient. Many within the profession naturally resist and resent this inversion, claiming that it weakens their ability (and power) to act in the patient's best interests.

As we have seen, when science and its province of meaning come together with Post-Enlightenment medicine, a search for certainty becomes a necessary component of the physician's way of knowing. Certainty here is not to be taken only as an addition to knowledge in the sense that it accomplishes the appropriation and the possession of knowledge. Rather, certainty is the authoritative mode of knowledge that is "truth." In the modern age, the scientific interpretation of knowledge becomes associated with truth, with certainty, and in this way makes a claim to power. The physician becomes entrenched within the borders of the scientific province and accent of meaning. The others who would critique his or her "profession" become adversaries out to subvert the logocentric value of his work.

As biomedical knowledge with its investment in techno-medicine has grown, however, the power differential between the physician and patient has gone beyond limits acceptable to the public. The *otherness* represented by the power and authority of the medical professional threatens people, and sparks a desire to assert some control over physicians through distrust, litigation, and legislative constraints [72]. This reversal reveals the slave's struggle against the master to be a struggle for mastery. Also, it shows the modern thirst for objective, certain, and clear knowledge. If we can know the facts, then we can know finally and fully; all disputes can be resolved, and chaos controlled. It is well to recognize that recent work on patient rights is a reaction to the profession's attempt to ensure control and authority (masked in name of public good). However, if this work merely inverts the power hierarchy, then trust will not be restored.

Instead, the two become locked in a competition that cannot be resolved and which becomes self-defeating. It is important to note that the physician is bound by this dynamic, too. This perpetual struggle is the reason that the application of moral principles such as autonomy, beneficence, or justice to health care issues does not find universal agreement and acceptance. Each party in the relationship interprets these principles in a different fashion, from within the boundaries of his or her own perspective. Each strives to reinforce the accent of meaning he or she brings to the relationship by excluding the possibility that meaning could be other than it appears to be. In the process of striving for the fully objective ego, however, each adversary denies the mutual bonds which tie them together.

This denial is bound to fail. The self is identified by what "it is not" as much as it is defined by what "it is." The effort to model the self after the "original" author of creation reveals a presupposition underlying the Western onto-theological tradition. The critical thrust of Western philosophy and theology debunks the metaphysical positivism that assumes a "direct relation exists between a sign and a corresponding object 'in reality'," and that this relation is discoverable through proper interpretation [118]. According to this positivism, if we can develop the correct method, we can know a phenomenon completely, beyond subjectivity and relativism.

However, the process of interpretation is one of continual re-creation formed in a tradition of hindsight. Every description of the physician or of the physician-patient relationship is an interpretation, an assessment that is built upon the criticism or destruction of its predecessor. No interpretation or diagnosis is final, and final interpretation is absent. The seeds of a replacement are sown when any "new" interpretation presents itself. In other words, interpretation begins as alienated interpretation. Once it becomes a paradigm, taken-for-granted, its destruction is assured. Therefore, there is no use seeking the presence of an original event or *text*. What is said, written, or done is important, but what is left unsaid, unwritten, or undone is equally, if not more, important in the potential that it represents ([155], p. 108). The meaning or value of an object is always contingent, never present in it. The interpreter's paradoxical task lies in the inability to finish the interpretation that he or she is compelled to make.

To summarize the theological analysis of trust-as-control, in an effort to usurp God's primacy humans have used the social institution of medicine and its social power to make health "present," to deny the irreducible dependency and interdependence that is part of the fabric of human existence. In the binary relationship between creator and creation, the predicate of divinity (wholeness, fully present, fully self-contained) is absent from us. Resentful and ashamed of our deficiency, we attempt to transfer this presence to the human. Our desire for the negation of uncertainty, illness, and death becomes the quest for certainty, health, and life. Such a movement creates and maintains the expectation that health is primary. This desire becomes total and domineering, since it can never be fulfilled. What is more, this quest can end only by a return to the origin of the search: the face of uncertainty and mortality mirrored in the image of a transcendent and absolute other-god. Confronted with this reality, the self flees again into a quest for fulfillment of need that results in self-denial.

The modern critique of theism transfers the attributes of God to humankind. Humanistic atheism then expresses a psychology of mastery in which self-assertion is a function of the attempt to negate the other in whose face the self sees itself reflected. This struggle for domination embodies the interrelated principles of utility and consumption that lie at the heart of technological consciousness ([154], pp. 64, 198). The psychology of mastery and the economy of domination represent efforts to deny death that can only end in narcissism and nihilism. This manner of self-assertion (or humanization) is finally self-defeating. The effort to master the self by mastering others (through the control of profession, the control of work, and technological control of the "world") theologically represents the attempt to dominate and subvert (the false) God ([267], pp. 13–14).

In a manner similar to other struggles for mastery, this drive to master alienated existence finally subverts itself. "The goal of this search is salvation – a cure that is supposed to bring health (*salus*) by closing the wound within. The search inevitably fails" ([267], p. 71). The hospital is a literal representation of the progressive humanization of the world. It is an effort to master "health (and) care" by organizing, building, and operating an environment conducive to the cure, control, or domination of illness and trauma. And yet, what could be more sterile than a hospital? Life (and death) in the hospital is the way in which many people have experienced most directly what it means to live without God in the world (see [235]).

Contrary to expectation, the repressive quest for domination and presence ends by disclosing the irreducibility of absence and the inevitability of death. Illness (as the absence of health) may be more important than ordinarily admitted in the philosophy or theology of medicine. The perception of mortality and the need for relationship is the void made present by illness; it is the wound that illness reveals. The effort to treat the illness can never resolve the dependency characterizing the human condition. The more we try, the more we exacerbate it.

Trust is never wholly or finally realized in social relationship. Maintaining its presence is a reciprocal and an endless task for the concerned parties. However, when trust is used as control, a self-assertive, atomistic individualism develops that is insufficient to maintain one's self in the face of an opaque destiny and visible sources of danger. A doctor who uses the illusion of trust sees himself or herself as acting autonomously; the patient who adopts this posture makes a similar claim. Both are working at cross purposes.

Trust requires another type of self-reference. It depends on risk. Risks emerge as components of decision and action. They do not exist by themselves. It is the internal calculation of external conditions that create risks. Trust is based on the circular relationship between risk and action, both being complementary requirements. Action is related to a particular risk as an external (or future) possibility, although risk at the same time is inherent in action and exists only if the actor chooses to trust ([169], p. 100).

In human interactions, consciousness is oriented to the appearance of an *other* in a field of awareness. The other is one who, like me, occupies a center of orientation to the world and its meanings. As a valuing being, the other presents himself or herself as a privileged locus of value, one not to be violated or simply used. In this way, the other sets limits to morally permissible projects. The other sets claims on me for loyalty and service. These claims have their basis in the distinctive being of the other as a center of valuation toward the world. Once a taken-for-granted accent is "shocked," if the other's claim is clarified, trust becomes possible. Trust, then, offers a means to transcend borders of multiple provinces of meaning, principally the two horizons of the physician and patient.

The concept of God as unmoved mover distances the deity from humanity. If God is envisioned as a "presence" whose divine life is enriched by relationship with humans, interdependence becomes the basis for relationship. A medical theological ethics must be fluid and historical, acknowledging the social nature of human existence.

NOTES

[1] Alan Keith-Lucas has written on the dynamics of giving and taking help, but his work does not deal directly with the peculiar condition of "dis-ease" in the physician's relationship with others. I will draw upon some of his ideas, but recast them from social work to a theological perspective; see [153]. More helpful from a theological point of view is Alastair Campbell's work [34, 35]. He exhorts us to resist making professionals into cult heroes and paradigms of loving concern, since this leads to the elevation of doctors as "gods." Such an elevation obscures the humanity and fallibility of the professional. Also helpful is Richard Titmus [273]. Also, see David Michael Levin [164].

[2] By "controlling presence" I mean a permanent influence sought or achieved by one member in which the other is reduced to an ontologically diminished status.

[3] This association of medical knowledge and clinical power, and the diminishment of the patient to the status of a body, reflects the "medical gaze" described by Foucault: "Doctor and patient are caught up in an ever-greater proximity, bound together, the doctor by an ever-more attentive, more insistent, more penetrating gaze, the patient by all the silent, irreplaceable qualities that, in him, betray – that is, reveal and conceal – the clearly ordered forms of the disease. . . . [T]o look in order to know, to show in order to teach, is not this a tacit form of violence, all the more abusive for its silence, upon a sick body that demands to be comforted, not displayed?" ([75], pp. 15–16, 84).

[4] There are multiple realities within any community. Within the Christian tradition, for example, there have been many interpretations of the nature of deity, each of which provided a center of meaning around which a church, sect, or denomination could form its identity. While two groups might define themselves over and against the particular interpretation of the other,

a feature common to each was attention to deity and some allegiance to the community of God.

[5] This God rules and is ruled by the nondialectical logic of simple negation: God is "not this," God is "not that." Norman O. Brown describes this logic as one of repression, since no positive statement is valid [(26], p. 161].

[6] This is the point made by Engelhardt's description of a crisis in values ([70], p. 3). For criticism of Engelhardt's theory, see ([133].

[7] Hartshorne and Reese object to this assertion, arguing that it is paradoxical for God to be all-inclusive, yet exclusive of suffering. To avoid the charge of patripassionism, they see Jesus as a loving and altruistically suffering human being who is not God, but a supreme symbol of deity, at the center of human suffering. This suggests that God is a being with absolute nonimmunity (not absolute immunity) to suffering ([116], pp. 162–163).

[8] Karl Barth appears to maintain the absolute gulf between God and mankind in terms of their respective "knowability" [(12], p. 174). The "search for the historical Jesus" represents the effort to find a more reciprocal relationship between the divine and the human (more convenient, Barth might say); see [114].

[9] Dependency and neediness, which arises from it, are fundamental components of human existence and cannot be denied. It is from this common condition that structures of meaning derive: "We are related to one another – existent being to existent being; value (positive or negative) is present in the fulfilling or crippling relations of being to being" ([202], p. 107). It also is from this common condition that interdependence and mutuality derive their importance in relationship.

[10] See Taylor's discussion of the principle of ownership in relation to utilitarian consumerism [267]. This acquisitive logic of need reflects the distinction between want and need in terms of health care. Because our need is inexhaustible, we want everything.

[11] By describing this sense of well-being and personal integrity, and by calling illness an "interruption" of this sense, I am not endorsing the World Health Organization's definition of health as "a state of complete physical, mental, and social well-being and not merely the absence of disease or infirmity." If this definition is accepted, it becomes virtually impossible for physicians, or anyone, to fulfill its implicit demand. However, for some authors, "holistic" concepts of health appear to be less misleading and more effective than biomedical concepts. For the WHO definition of health, see ([13], p. 48).

[12] The humanistic tradition has never realized the deep rooted need within people to destroy. Attempts to establish "good" within the world, if motivated by shame, can release tremendous destructive tendencies.

[13] See [88]. The roots of the struggle between deontological and consequentialist ethicists over which theory will provide maximum individual freedom lie in the drive to relieve human estrangement. Also, see the discussion of the adverse influence of atomistic individualism on a sense of commitment, in [15].

[14] Nietzsche's refusal to "bow worshipfully before the saint" led him to an idealization of egocentric selfishness and the rejection of self-annihilation ([206], p. 65).

[15] Freire discusses the desire of the oppressed workers to reverse their positions and dominate their oppressors. However, as he points out, this reversal only perpetuates the economy of domination [88].

CHAPTER 6

TRUST-AS-FAITH: GIVING AND RECEIVING IN DEPENDENCY

The economy of domination within social relationships that uses trust-as-control is caused by the denial of mutual dependency. The growth of the scientific structure in medicine reflects a desire to minimize uncertainty and to control the limits of medical work. This desire has shaped the profession's sense of its identity, the value of its work, and consequently, the nature of its relations with the public. While this orientation has produced great advances in medical power and authority, it also structures relations among people in particular ways. The patterns of interactions among people become more objectified, more routinized. This structure has created a "distance" between the vulnerable patient and the powerful physician. People have accepted the arrangement – have come to take it for granted – in return for medicine's promise of help in times of need. In the face of medical knowledge and power that they cannot really comprehend, the public (and the physician) must assume that the fiduciary component of the profession's identity places the patient's interests above the physician's own self-interest. Both parties are bound to a fiduciary relationship, yet one that has become a peculiar institution.

The promise of biomedical science to overcome illness and disease is premature. The investment of huge sums of money and concentrated efforts in biomedicine have led to enhanced medical knowledge, some undeniable successes in treatment, and technological advances. However, the presence of illness and uncertainty beyond the ability of physicians to control reveals the fact of human limitations. As people become ill, their human condition of dependency and vulnerability is laid bare. Patients and families facing the interruption of illness naturally want some sense of control in their lives. The physician wants to control and minimize uncertainty and contingency in his or her work. Trust-as-control provides a measure of satisfaction for each party.

After all, as a social dynamic trust-as-control allows some grounds for interaction. It reinforces the physician's "authority" in medical matters ([123], pp. 39–62), and it encourages a commitment to patient autonomy. It may support a person's confidence in physicians' skill and knowledge, and it may provide a sense of self-confidence to physicians. This approach to relationship, however, represents a negative movement. It does not foster a depth of trust between the parties in the medical relationship, i.e., that each is working to mutually enhance the other's well-being. With trust-as-control each person holds back full commitment and approaches the relationship with caution and reservation. Trust-as-control has moral, social, medical, and theological repercussions. It adds fuel to the feeling of moral impasse in matters of medical ethics. Socially, it reveals the presence of interest groups. Medically, it fosters the practice of defensive medicine. From a theological perspective, it also does damage to the community of interdependent people by fragmenting a commitment to mutual enhancement.

By showing that trust-as-faith suggests a theological reorientation of this uneasy and precarious balance of power, I propose a constructive way out of this impasse: Trust-as-faith transforms the economy of domination into a free exchange of dependency and vulnerability. In discussing the Christian view of faith as a gift of God and faith as the human response to that gift, I concentrate on the following points: faith (*fides*) is God's gift offered to restore the proper relationship between God and the world. Faith (*fiducia*) is the response to this gift. In this giving and receiving of faith, there is a transformation from an alienated existence to one of faith-full trust. The "contextual" and reflexive nature of trust for those involved in medicine is compatible with these theological concerns.

There has been a shift from theologies of medicine to a focus on medical theologies. Before the last two decades, the preponderance of writers on Christian theology and biomedical issues assumed that the question was: What does Christianity (the Bible, the tradition, and the church's ecclesiastical authority) have to say about the ethical issues raised by medicine? Now the question has become: What does the phenomenological experience of human health and illness have to say about perceptions of faith – of the experience of God and of interpretations of Scripture and tradition? In part, this movement is a recovery of 19th century liberal theology's emphasis on experience as important theological data – that insight into the ways of God can be gained in dialogical or connexional human relationships ([200], pp. 187–190; [266], pp. 125–130). The experience of otherness (the awareness that things can be other than previously assumed, and that the *self* is not necessarily privileged), especially in the interruption represented by illness, is very uncomfortable. In the face of this experience, faith as fiducia, or trustful response, becomes the theological basis of the fiduciary component of social relationship. It brackets the distrust and alienation from otherness characterizing postmodern experience. However, otherness cannot be eliminated. A medical theology begins, therefore, with a recognition of the necessity of trust "in spite of" the irreducibility of human dependency.

The chapter begins, first, with the description of a theology of dependency based on the human condition of incompleteness. Dependency is an inescapable part of human life in that people rely on others, on their community and their culture, for meaningful existence. People may act upon these meaning structures given to them, but their actions themselves are interpreted according to the present form of meaning. Thus, people are not self-sufficient, complete beings. Each person receives his or her life from beyond himself or herself. In theological language, "to rely upon others, and ultimately to rely on God is to experience the limits of self-sufficiency, and thus to recognize the finitude of oneself, others, and all that is" [112]. The drive to dominate and deny dependency, however, causes people to seek independence and self-sufficiency. Traditionally, within Christian theology the inability to overcome this condition reflects the "fallen" condition of human existence.

The incompleteness of humanity finds completion in reliance upon and faithful relationship with God. This completion does not eliminate dependency. Illness, for instance, is a time when we are compelled by necessity to realize that our lives are not our own. One's embodied existence, which was taken for granted before, and the expectations for the continuation of that reality are called into question by illness. The condition of illness represents a situation in which one's ordinary experience of everyday reality is bracketed.

The problems of dependency in the context of medicine are moral ones. In illness, the patient is immediately aware of his or her vulnerability and clings to the hope that the physician is trustworthy. At the same time, dependent upon others for help, patients are exposed to the threat of domination and control by the physician's knowledge and power. Illness as a representation of dependency is an experience which calls into question our everyday ideas, expectations, and values. Theology works to place these concerns into another context, the locus of which is beyond human control.

A second reason for theological reflection is that most secular philosophies, other than existentialism, do not begin with human frailty. They build on the strengths of human reason without appeal to God, or upon our mastery of nature, to the neglect of human weakness. Theology looks squarely at the depths of being human, dependency and all. Thus, in the last chapter I offered a theological examination of dependency and the use of trust-as-control. In this chapter, I discuss the theological implications of dependency for the attitudes and dispositions with which physicians' relate in a fiduciary fashion with patients.

TOWARDS A MEDICAL THEOLOGY OF TRUSTFUL DEPENDENCE

A medical theology based upon recognition of mutual dependency begins with the perception of the social basis of human existence. Even where people claim autonomy in their actions, some sort of outside assistance makes that independent life possible. An individual from birth is supported by family, society, culture – his or her life is given meaning from the social matrix into which he is born and in which he grows up. His understanding of life, his values are gained from interaction with others within an environment that existed before his birth. Choices involve weighing decisions between alternatives, the existence of which are present because of social structures of meaning.

The fact of human dependency has implications for physicians' expectations of their work with others. Familiarity, confidence, and trust are different modes of asserting expectations – they represent different types of self-assurance. Belonging to the same family, familiarity, confidence, and trust seem to depend on each other and are capable of replacing each other to some extent ([169], pp. 94–108). Confidence and trust have been discussed. Familiarity draws a distinction between familiar and unfamiliar fields and places a priority on the familiar. The unfamiliar remains opaque. Ordinarily, there is no need for conscious self-reflection on this distinction: One is familiar, not unfamiliar, with oneself. Illness, however, represents a moment in which a person becomes aware of the intersection between the "unfamiliar" and the "familiar."[1]

As Niklas Luhman points out, trust has to be achieved in a familiar world, and when changes occur in familiar features of the world there is an impact on the trust in human relations ([169], p. 95). Phenomenologically, illness transforms the taken-for-granted world of everyday life into one unfamiliar in its dimensions. Human society has developed meaning structures and symbols to make this unfamiliar interruption familiar. In spite of these structures, however, there remains something irreducible about the experience of infirmity.

Typically, physicians and patients seek a familiar set of relations because they have confidence in the symbols that orient them to the familiar roles of the medical relationship – they seek a way never to leave the familiar world. During the transition from the unfamiliar to the familiar, during the shock that accompanies illness, they seek to maximize what is familiar to them. On one level of meaning, the experience of illness is controlled by a complex process of domestication. At the same time, there is something about illness that cannot be domesticated or eliminated: the recognition of dependency and vulnerability.

During such moments of disruption, a person perceives the risk inherent in social interaction – the riskier the moment, the closer to an acknowledgment of dependency, to awareness of the "thrownness" of existence described by Martin Heidegger ([127], p. 134). Threatened by that which is beyond human control (the givenness in the social and physical world), to deny the risk to the self of one's dependence on anyone and anything beyond the self, the person become possessive of his or her identity and suspicious of others. This possessiveness disrupts the free and open exchange of relationship – in a sense, people clutch their dependency to themselves. Rather than accept the human condition in all aspects as a gift, people attempt to eliminate their dependency and neediness. They, therefore, cannot give easily and cannot receive easily in turn. Trust as a medium of social exchange becomes a commodity to be bartered. Dependency and the feelings of neediness it fosters come to be perceived as weaknesses which people are afraid to reveal, yet these conditions lie at the root of existence [271]. The effort to deny dependency can lead to the false expectation that the needs of everyone can be eliminated. This egotistical view of the ideal of service is a distortion of the giving and receiving relationship and becomes the conceit of philanthropy when it is assumed that one's commitment to his or her fellow human is a gratuitous, rather than a responsive or reciprocal, act ([175], p. 70).[2]

The drive to eliminate dependency provides the conditions for the "sin" of pride.[3] In the condition of estrangement from God, the gifts of the other (which include the gift of his or her needs) are reduced to "objects." In doing so, people refuse to acknowledge and express their own vulnerabilities and dependency; they objectify themselves in turn. Domination and exploitation become motivating forces in social interaction as people attempt to mask their neediness and dependency from others and from themselves. As a result, they are unable to accept life as a gift and cannot return that gift to the world with authentic gratitude.

All people are dependent in some way. Aware of their finitude, they feel a need for a long and healthy life. Aware of their individuality and sense of self, they feel the need for social relationships. One of the great human fears is suffering alone, of being abandoned by others in moments of need. Hauerwas argues that in the Christian tradition, therefore, the church is to be a people who are faithful to one another by willingness to be present to one another despite human vulnerabilities [1[illegible]]. Precisely because humans are dependent creatures, there is a trust-as-faith given to and received by others who are willing to be present to another in the midst of their vulnerability, neediness, and incompleteness. This willingness to be present suggests that the appropriate model of medical relationship is trust between mutually dependent parties. Trust-as-faith is the quality of relationship that enables partners to freely give and freely receive in mutually enhancing, enlivening (healthy) ways.

FIDES AND FIDUCIA: THE IMPORTANCE OF TRUST-AS-FAITH FOR THE MEDICAL PROFESSION

According to Gustafson, the first task in order of importance for theological ethics is to establish convictions about God and God's relations to the world, to make a case for how some things "really and ultimately are" ([113], p. 98). This task seeks to show the manner in which faith sheds "new light" on the fullest and most profound dimension of human existence [178]. Driven by a need to deny his or her dependency and the awareness of his or her own face in the sick bed, the illusion of trust threatens relationship with the patient whose presence paradoxically challenges and confirms the physician's accent of meaning. The task for theological ethics is to reorient this distorted exchange of relationship and to restore the sense of trustworthiness and trustfulness for both the physician and the patient. If a common understanding of the nature of trust is eroding, a theological perspective offers an understanding of human dependency and interconnection which restores the epoché of doubt and despair that is disrupted by illness.

In the Christian tradition, it is important to remember that alienation from God and self-estrangement characterize the sinful nature of the human condition. "Man is sinner," and sin implies the attempt to be absolutely independent and autonomous ([12], p. 529). Since, however, a person's powers and abilities to know beyond the moment are limited, when one makes himself or herself the object of self-consciousness, he or she discovers that the actual self always falls short of the ideal self. As a result, to deny his or her inadequacy and incomplete nature, one is driven to self-aggrandizement and becomes alienated from his or her creator and one's self in the drive toward an economy of domination of the other [204].

Consciousness of this drive becomes a necessary pre-condition for restoration of wholeness. In the postmodern world, in the culture of criticism that is fostering a climate of suspicion between physician and patient, to retreat into a protestation of codes, professional ethics, or medical knowledge and skill creates the illusion of trustworthiness. Following the Augustinian view that sin distorts all human faculties, one cannot know the truth about himself or herself and God until one's understanding and will have been transformed. The person cannot free himself or herself from bondage to sin, from the drive to domination, by his or her own will. It is only through God's revelation of himself in the Incarnation that people become aware of the depth of sin, and only then does faith become possible as a grateful act of willing response to God's grace-filled gift ([47], p. 400).

Faith does not free the person to be free, independent, or completely autonomous. Rather, it frees the person to be free in interdependence, in the recognition of a mutual dependency upon others or God in a relationship that changes over time, and calls faith into question. Consequently, from a theological point of view, the profession cannot control the limits of its work in any final fashion, and the attempt to do so leads only to the illusion of care and concern. The embodied presence of the other person who establishes the limits of the physician's power represents a gift that calls for a response. The danger in the unique social relation that medicine represents is that this exchange of gifts may become distorted if one or the other places himself first in an effort to dominate the relationship. The physician must work to bring out the content of and the confession of mutual dependence and trust, to help the patient restore a frame of reference providing a meaningful existence. Otherwise, medicine becomes another type of domination in which trust in relationship is a commodity to be bartered between self-interested parties.

This bartering economy that leads to domination is grounded in the anxiety caused by the perception of human finitude. The natural attitude works to bestow upon everyday reality a sense of permanence, of explanation. In the moment of phenomenological disruption, the natural attitude is bracketed and the risk of relationship (the awareness of mutual dependence) is revealed. Trust-as-faith restores the natural attitude in which life, health, and relationship again assume a "reality," and yet no longer are taken for granted. This seeming paradox is due to the hidden, yet revealed nature of God.[4]

According to this theological perspective, with the acceptance of this gift, freely received in faith and in trusting obedience to what has already been given, humans are free to be in relation to God and others – personally, not as types. At this moment, the illusion of temporal existence and power is called into question and revealed in its true nature: The structures of power formerly taken for granted as complete in themselves are shown to be incomplete, dependent, and truly interdependent. A person is revealed as a creature whose willful pride covers a refusal to recognize dependency and need in the drive for control and domination. This refusal distances him or her from his or her true self as a creature of God, truly dependent upon others in the world. Society and the human do not stand apart from God and each other but are fully realized and completed in God and in each other.

These are not relations that humans can create on their own out of their dependency and need. They result from the renunciation of self-determination and self-sufficiency, from the realization that the desire for self-control is illusory. Trust as expectation of the beneficence of others can not be rooted in self-possession or control but depends wholly upon a free and mutual exchange of neediness.

The human condition is embodied in lived existence and social reality. The recognition of dependency and the neediness it creates arouses anguish because for much of our lives we ignore the neediness that links us to others. For the physician, particularly, engaged in working with those who are ill, who are suffering and vulnerable, the medical accent of meaning provides structure that insulates the person from this recognition. Through the clinical perspective, through organizational structure and the dedication to knowledge and expertise, and especially through the profession of altruistic service to those in need, the physician seeks to control the uncertainty of illness that threatens the order and meaning that medicine represents. Relations with the ill, however, are inescapably ambiguous and filled with tension [266]. Through contact with illness, whether directly in a medical encounter or indirectly through research and teaching, the physician represents one side of a relationship whose existence is due to a concrete manifestation of dependency and neediness. By knowing more, by trying one more therapy, eventually by manipulating and controlling people, the physician attempts to convince himself or herself of his or her freedom and mastery of the world, avoiding admitting how much he or she receives from others.

From the standpoint of Christian faith, this avoidance of dependency is mistaken. The natural desire to deny our dependent nature creates an economy of domination that does not eliminate needs but only masks them, leading to further efforts of denial. The recognition of our contingency does not mean we are without an absolute. As H. Richard Niebuhr writes,

> In the presence of their relativities men seem to have three possibilities: they can become nihilists and consistent skeptics who affirm that nothing can be relied upon; or they can flee to the authority of some relative position, affirming that . . . a value, like that of life for the self, is absolute; or they can accept their relativities with faith in the infinite Absolute to whom all their relative views, values and duties are subject ([201], p. 238).

With faith in this Absolute, Drew Christiansen argues, it is possible to let go of a desire for complete self-control, to accept support from others, and to find fulfillment in a community of care [44]. Overcoming the economy of domination produced by the use of trust-as-control involves placing one's basic trust in a presence that is beyond one's control.

Being able to rely on the presence of a physician, with the trust that he or she is present for the person's benefit, sustains patients at a time when they are most vulnerable. However, as people who are shaped by the meaning structures of their profession, physicians have a dual loyalty to their profession and to their patients. Certainly, fidelity to patients rebounds to the benefit of the profession – it serves to maintain public confidence in the physician. However, loyalty to the profession or to patients can become self-interested. If the physician begins to consider the other only in his or her value-relations to himself or herself, the giving and receiving of presence between them will be twisted into trust-as-control. By accepting in trust the gift of the absolute faithfulness and trustworthiness of God, it becomes less difficult to accept dependency and vulnerability ([201], pp. 249–252). The physician who is able to receive in faith the presence of others can return the gift of presence to others. Paradoxically, from the Christian point of view, recognition in faith of human dependency establishes the person's sufficiency to live in the face of that condition and frees the person to the satisfaction of faithful giving and receiving. Therefore, May can argue that ethics grounded in the covenantal relationship with God enables one to be responsive and requires one to be available to the other above and beyond the measure of self-interest ([175], pp. 73, 74).

Taking into account the fact of historical change and the limitations and relativism it contains, however, how can the theologian locate any absolute moral norms? How can the theologian establish an appreciation of and interpretation of Christianity and culture that allows for the autonomy of both and yet allows the two to be interdependent as well? Since the insights of theology, as well as other sciences, change over time, there is a need within theology to move beyond itself, even as it is being developed. The pronouncements of a theology cannot be absolute; theology can function as a form for consistency, but it cannot be final. Thus, the essence of Christianity must be "recreated" in every age ([281], pp. 124–181).

Therefore, in developing an argument for the place of theological ethics in biomedical matters, structures of sociality and human intersubjectivity constitute meaning and values for people, a way of seeing the world that orients them to a lived reality. Unfortunately, as this orientation becomes "taken for granted" and objectively abstracted, we can lose the awareness that values are not isolated or fixed entities, existing outside of encounter with others, but emerge in relations among beings. Value does, then, have objective presence – it is disclosed, rather than projected, in the encounter of being with being [132].

Loyalty emerges in relations among beings, but because it emerges in encounters that transcend limited forms of human need and awareness, it is not "subjective" in a dualistic sense. In the deconstructionist critique, "subjectivity" refers to an enclosure of the self that, by reducing reality to the reality of the self, projects loyalty as its self-centered attentions dictate. Trust-as-control is a reflection of this subjectivity.

A Christian theological ethics finds the locus of value in God. The God of radical monotheism, however, "shocks" the relation of being and value because God lies beyond both being and value, and "one cannot say that God has need of any being external to himself" ([202], pp. 112–115). According to the deconstructionist critique of traditional notions of God, as a quality of divinity "separateness" becomes extolled as a virtue in and of itself. In this logocentric schema, relation to otherness implies self-deficiency and fosters the control of otherness for the sake of the separate self. This perspective is informed by the paradigm of the *otherness* of God, the model of a Person who, having no needs, stands apart from the community of being. Perhaps God's isolation is a projection of the human need to maintain power by controlling dependency rather than the presence of an inclusive love for that which is in need ([132], pp. 89], 92).

For Niebuhr, "self-relatedness" has no meaning apart from "other-relatedness" ([202], p. 105). If we are to find the presence of the transcendent God, we need to reimagine or reinterpret the metaphor of *imago Dei*. God, like us, has needs, and it is through the giving and receiving of trust, one in the other, that God's presence is known through the interdependence of everyone. For the physician, the values of knowledge and trust, a dedication to expertise and a fiduciary commitment, shape his or her self-identity and the "limits" of medical work. The metaphor of trust-as-faith reveals the worth, relatedness, and need that all people share in making meaning in their world. Although this metaphor may not be the conclusive result that would enable us to say, "This is the Christian answer" to the question about the contribution of theological ethics to bioethics, in the faith that has its investment beyond relativity, in the pluralism of relatedness, there is a freedom of interdependence in moral decision-making.

IMPLICATIONS FOR A MEDICAL THEOLOGICAL ETHICS

The study of the medical profession's commitment to knowledge and a fiduciary relationship with others, the way in which these foci shape and are shaped by the physician's accent of meaning, highlights several issues that have a theological resonance. The use of trust-as-control in the medical relationship, and the economy of domination that results, echoes theological concerns with alienation in human relations. The resulting theological discussion of faith and trustful dependency brings us to the implications for a medical theological ethics.

A Phenomenology of Illness

A medical theological ethics must recognize the importance of a phenomenological approach to the medical encounter. The experiences of physician and patient are uniquely constituted [275, 295]. Michael Gordy asserts that bioethical issues "presuppose a social nexus and have to be discussed in terms of the social relations between people" [101]. Much has been written about the need for social contract or covenant in medical relations.[5] James Sellers wonders if the "we-relation" does not precede either of these models. Without limiting the options of contract or covenant, the "we-relation" sets out what is prior: a sense of "we-ness" that takes us across time, "growing older together," in the pursuit of a common understanding of being human ([248], pp. 9–10).[6] Thus, an attempt to interpret the meaning of trust and trustworthiness in the structure of medical work must be conditional.

Just as the use of traits or function to define the medical profession presupposes an already existing idea of profession, how can we trust someone or something when, phenomenologically, our trust can never be "final." Our expectations of others, our willingness to trust them in situations of risk, are based on our biographical situations and intentionalities, on past experience, present understanding, and anticipation of future needs. Paradoxically, trust can only be momentary, yet we are called upon to trust beyond the moment. Since the "fullness" of trust can be present only when absent, complete certainty is always already lost. In other words, as Mark Taylor puts it, "there is always a serpent in the garden."[7] If we are uneasy with the status of trust in relation to physicians, the effort to make trust fully present, to guarantee trustworthiness, already represents the turn to self-concern and self-deception in the attempt to recover complete certainty: We already "know" that physicians are fallible; we suspect that they are untrustworthy. As a result, deficiency or dependency is believed to be the result of a fall, a breach or break, a condition requiring control. Thus, to be ill is not merely to realize one's dependence upon others; it is to be ashamed and resentful that one cannot control the contingent nature of one's existence. In the quest for control, because of it, we lose sight of the conditional, momentary nature of trust. Since certainty is always already lost, our perception of contingency and finitude grows, and trust-as-control becomes the mode of interaction.

In this mode, people long for secure selfhood, yet they are unable to secure it. For instance, in the attempt to secure the self against invasion, people look for "territory" where they can be secure and powerful. This militaristic metaphor conveys an image of the self as finite and vulnerable, able to be overwhelmed if left unprotected. This concern leads to a preoccupation with one's own independence, as commitments or alliances with others always limit or bind a self and increase its vulnerability. For the self struggling to be independent, the only real powers beyond the self that are acknowledged are powers that are seen as a threat, and the only kinds of relationships with other persons that are comfortable are those that are distant and superficial, or those in which the self is in charge – as parent, teacher, doctor ([258], p. 277).

However, the self also can respond to the interruption of illness by acknowledging the interdependence of people, that the person, though finite and limited, is a whole who is a part of the whole. With this response there is an awareness of mutuality in relation, that the giving and receiving of relationship is not a defensive battle in which one wins as the other loses, but rather an exchange in which self-benefit is derived from some attention to the benefit of the other. This response suggests one basis for a medical theological ethics.

With the focus on human interdependency, there is a shift from a theological understanding of medical ethical problems as a matter of wrong acts to an understanding of "illness" as alienation from our intended personal and communal relationship. This shift eases the strain that occurs when theological ethics and philosophical ethics become involved in a shoving match over "first principles." Theological ethics recognizes that sin is fundamental alienation from our divinely intended relations with God and with each other. Sin, therefore, is not fundamentally an act, but rather the condition of alienation or estrangement out of which harmful acts may arise. Such alienation from the way God intends us to be may give rise to harmful acts, but the act is rooted in the distrust and suspicion of the prior condition. Sin lies in the alienation caused by a dualistic attitude that sees the body as an object that "acts" in the physical world but is only made meaningful in terms of its superficial actions (which to be truly meaningful must be constrained by the higher spirit or purpose). This dualistic attitude leads to a denigration of dependency, and the embodied condition becomes "evil." In this habit of mind, "illness" and "death" become invaders to be combatted. Metaphorically, the sick person's body signifies a war zone in which the battle is fought [236, 260]. Winning or losing becomes the focus for the physician and the public, and the person, patient or physician, is lost in the struggle.

The focus of the medical profession on the acquisition and utilization of knowledge to provide "health" care, to rescue the body in distress, follows from the inescapable dependency of our physical lives. However, such a clinical perspective pushes physicians towards an attitude of mastery and control of the environment (e.g., the body in the bed). As long as the motivation for this attitude lies in the denial of mutual dependency, there will be a disregard for the conditions that move in the direction of human wholeness, be this in individual health, in social relations, or in spirituality.

Piero Camporesi claims that sanctification means growth in holiness (or wholeness and health – the root word is the same) ([36], p. 221). It may be that God intends increasing wholeness to be part of human redemption. As with any other belief, this view can be distorted, but it need not be. In medical matters, sanctification can mean growth in acceptance of one's physical existence as well as one's spiritual self-acceptance. As the epoché of the natural attitude reflects trust-as-faith, the person's sense of self and community, in relation to God, may be more fulfilling. A phenomenological interpretation of trust's effect on a physician's "reality" serves to deepen the doctor's appreciation of the relational dimension of his or her work.

Embodiment

In developing a medical theological ethics, with an appreciation of the phenomenological approach to medical work, there is a shift from understanding embodiment as either incidental or detrimental to the experience of God, toward an understanding of embodiment as intrinsic to the divine-human experience. A Cartesian dualism has marked much of the profession's perspective on health care ([295], pp. 187–190). In this dualism, spirit is opposed to body, with spirit assumed to be higher and superior and the body lower and inferior. In the philosophy of medicine such dualism presents health as superior to illness. The companion of this dualism has been the association of the gatekeepers of health with the spirit and the patient with the body.

From the phenomenological perspective, such dualism suggests a denial of the suspicion that the body is not fully self-sufficient [91]. As ontological priority is given to the spirit, the body becomes discounted, something to be denied or dominated. Health becomes a priority, and the gatekeeper to health assumes a controlling position over and against the *body*. An appreciation of the importance of embodiment shifts this image. I *am* my body; yet in another sense I am not *just* my body. "The relationship of person to embodying organism is more complex: not only 'mineness' but also radical *otherness* is inherent in it" ([294], pp. 188–189). In health, I take my body for granted; in illness, I become aware of its otherness. With a phenomenological appreciation of the meaning structures that constitute perceptions of reality, health and illness become more complexly related: The presence of one does not remove or erase the "presence" of the other.

A realization of one's embodied state does not erase forever any sense of well-being or health. The natural attitude works to reconcile any dissonant concern within the person's frame of reference. If the natural attitude denies recognition of embodied vulnerability, the body's presence merely is hidden. People do recover from illnesses and resume a sense of health in their everyday lives. For some people such a recovery is easier than it is for others: The person recovering from the feeling of weakness caused by the flu may resume the activities he or she previously enjoyed more easily and fully than one recovering from a myocardial infarction or cancer [80]. However, the recovered natural attitude can bracket an awareness of the "lived-body" experience. One may regain confidence in one's body and return to a prominent sense of inviolability. This movement results in a suspension of doubt in the person's bodily dependence, but it also interferes with an appreciation of his or her embodied self. The gift of the body, its "givenness," is not always appreciated.

One result of this skewed ontological position has been the assignation of positive value to the physician and a negative value to the patient. The physician is associated with health and an image of self-sufficient power, while the patient (the person with the problem) is somehow deficient, less than autonomous and self-possessed – one who needs the physician and is dependent upon the profession's knowledge and powers. The physician's presence then becomes primary, and the recognition of the patient as a mutual partner in medical work is absent. Recent attempts to invert this relation only serve to replace the priority of the physician with that of the patient; the slave replaces the master as master, and the economy of domination continues [88]. We must look beyond guarantees of autonomy and informed consent in medical care to find a renewal of the fiducial in health care. A review of Christian notions of divine and human relation points the way.

Implicit in the spirit-body dualism has been the notion of divine centrality, impassivity, and transcendence. If the physical is marked by decay and illness, and if the spirit is perfect and healthy, then illness has no connection with the divine other than as retribution for human actions and transgressions [253]. Those who are ill, or dispossessed, or socially unacceptable are relegated to the periphery of society. Since God does not get sick and is without need, human dependency and vulnerability, of which illness is an example, is allowed to have no real connection with our experience of God.

If we are to have a theological understanding of the role of trust in physicians' identity, we must renew an appreciation of incarnational theology. This theology emphasizes that the most decisive experience of God is not in doctrine but in the Word made flesh – and in the Word still becoming flesh.[8] Spiritualistic dualism has conditioned so much of our response to the physical body that a strong theme in Christianity is the disembodied notion of salvation: Salvation means release from the lower (fleshly) into the higher (spiritual) life. However, in an incarnational faith salvation embraces the redemption of alienated bodies as well as other estranged dimensions of our lives. Grace signifies God's unconditional, unmerited acceptance of the whole person. As a free gift, the possible becomes actual. God's free gift as God becomes present among us (in the person of Christ-who-suffers with us).

In a consideration of the application of trust-as-faith to the patient-physician relationship, we must be careful not to stray too far from the practical experience of embodiment. When one is ill, the embodied context of relationship must be reappreciated. The phenomenological quality of health and illness returns us to the discussion of the moral nature of the physician-patient relationship. The person experiencing (popularly assumed to be "suffering") the ontological crisis of illness and the vulnerable existential condition that accompanies it does not have the knowledge or skill to cure his or her own bodily or mental illness, or to gain relief from pain or anxiety. His freedom to act as a person is compromised. Unless the person denies or ignores this condition, he or she must seek the help of another and must occupy a vulnerable relationship with the professional who possesses what the patient lacks ([222], p. 159). Theology gives new attention to the insight that a more comprehensive approach to health care is crucial to God's design that we not suffer in isolation and loneliness, but in communion and community.

Historical and Social Focus

An appreciation of trust-as-faith undermines an understanding of the medical profession as ahistorical and reveals it to be an historical enterprise. At the same time, there is a shift from understanding illness and health care as private and individualistic issues to understanding them as public and social ones. Illness will always be deeply personal, but personal does not mean private. On the agenda of policy makers today are social justice issues regarding access to health care. In an era of scarce economic resources and a demand for higher utilization of health care, the position of the physician in health care will be inescapably public. Also, the medicalization of society (the involvement of medical professionals in areas outside of the strict purview of their field) ironically has contributed to the public's sense of involvement in health care matters. As a result, the image of authority and power associated with the profession is changing. The physician must operate under public scrutiny in a more obvious fashion. This development has a reciprocal effect upon the physician's self-image, identity, and understanding of his or her work.

A sociological and phenomenological study has shown that knowledge and fiduciary commitment are defining foci of professional status. While it enhances the physicians' ability to function in the modern clinical setting, trust is not an essential attribute of physicians. Trustfulness and trustworthiness are essential for the practice of the medical profession. Phenomenological bracketing shows that "changeless" perceptions and understandings are not changeless; fixed structures of meaning have a way of being interrupted. Once we appreciate the historical and social nature of trust, we can say that trust is a virtue because it necessarily enables people to overcome shocks to and allows the reintegration of human relationship, over time, under fluid conditions. Trust itself is not an absolute condition; blind or unquestioning trust distorts relationship as much as skepticism. As distrust becomes tinged with the economy of domination, it begins to acquire an element of alienation even as it structures human relationship and, consequently, the individual's image of him or herself and others. Trust-as-faith, however, provides a critical appreciation of human frailty and susceptibility to the economy of domination as well as the willingness to give one's presence to another in relationship.

A Return to "Practice"

There is another aspect of the turn to a medical theological ethics. If the horizon of meaning of the phenomenology of illness is adopted, illness is that aspect of human existence that calls into question a person's sense of personal and communal stability. What formerly seemed to be so dependable suddenly is abrogated. The dominant conception of bioethics encourages the turn to trust-as-control. By appealing to universal principles applied to particular cases, we too often fail to realize the importance of concretely lived experience that comprises the actual work of physicians and patients wrestling with complex concerns, anxieties, goals. We do not have abstract existences as moral agents within the clinical encounter – there is too much at stake for all concerned.

The taken-for-granted view of ethics commonly at play in the clinical setting does not respect the personal integrity of the concrete moral subject. It requires that people give up their personal points of view in exchange for a universal and impartial point of view. This is absurd since the person is requested to give up the very thing that constitutes his or her personal identity and integrity. The community, the particular social group to which we belong, is usually at the center of our moral experience. Varying views of community as the focus of moral life can be found in the work of thinkers such as Alasdair MacIntyre, Martin Buber, and Stanley Hauerwas [28, 120, 183]. Each of these scholars concentrates in his own way on the communal interdependence of persons that underscores the moral life.

There are two essential elements in the concept of "community." The first is that there is a common purpose or common center around which people gather and which supports and extends their reason for being.[9] Hauerwas states that, "a community is a group of persons who share a history and whose common interpretations about that history provide the basis for common actions" ([120], p. 60). The members of the Christian community, for instance, may be led to worship, act nonviolently, or become involved in AIDS work through their common understanding of the gospel message. The nature of this relationship is best seen as one of commitment to the common purpose. Writing of "practice," MacIntyre stresses the coherent and complex form of socially established cooperative human activity through which goods internal to that form of activity are realized in the course of trying to achieve those standards of excellence that are appropriate to, and partially definitive of, that form of activity, with the result that human powers to achieve excellence, and human conceptions of the ends and goods involved, are systematically extended ([183], p. 14). If the participants lose this commitment to the standards that define the profession, this tension with the larger society is weakened, and the internal goods that help define excellence will erode: the community will fragment from internal and external pressures (see [180]).

An exception to this circumstance highlights the second feature of community. If persons lose their commitment to the common center, their community may still function if they retain their commitment to each other. The other central feature of community lies in the common bonds of its members.[10] In the struggle to retain commitment to each other, the original purpose may become modified or reinterpreted in order to accommodate the changing perspectives of the members.

Any viable community such as the medical profession will reflect these two general features. Its members to some extent will be committed to a common purpose or center and to one another. Insofar as community is sought as an end, perhaps because of its general value for human good, there are two ways it can be pursued – by attaining agreement on ends or by promoting the sorts of bonds between people that are necessary for the community to exist. Given the pluralistic culture and the fragmentation of society into "atomistic individualism," agreement on a common end or purpose to which all can agree is not likely [15]. The other possibility is to foster the relevant commitment between people. If commitment to a common center (metanarratives such as God or the taken-for-granted trustworthiness of the medical profession) is lost, if commitment to each other is moving toward an atomistic individualism, on what do we ground moral relations?

One of the appeals of this approach to a discussion of theology and bioethics lies in its assumption that community is one essential aspect of human life and in its emphasis upon the relationships between people rather than on a prior debate about the "rights" or "autonomy" of individuals ([61], pp. 190–191). Rights, duties, obligation, or moral principles such as autonomy, beneficence, and justice are grounded in the fact of human community. A focus on the commitment between people has the virtue of underscoring the need to appreciate the historical and social structure, the mutuality of relationship, and an economy of authority and power in human relations that are the bases of moral life.

To repeat, the two central features of community are commitment to a common end or purpose and a commitment to other members. One crucial aspect of these commitments, that is, an outgrowth of the essential bonds that make a community a community, lies in the willingness of persons to be present to one another in times of suffering and need. Presence is essentially a matter of availability to and caring for the other. Its opposite is captured in the concepts of indifference, abandonment, and desertion. When persons are suffering or in need, and are left to their own resources, the community necessarily fragments. This commitment to be present to one another is correlative to the commitment to a center. If one commitment is weakened, the other begins to erode. If one is reinterpreted or modified, the other will change as well.

In the postmodern period, commitment to a broadly based center of community is problematical. Under a strong postmodernist critique, claims to meaning fall to chaotic "shape-shifting" [79]. In these conditions, it is no wonder that fragmentation of mutual concern and presence follows. As the taken-for-granted understanding of a common center slips away, there is movement towards atomistic individualism; why not be self-seeking, if there is no real meaning beyond the momentary? For people to live in community, they must be present to each other. As their common center, end, or purpose erodes and becomes absent, however, their appreciation of mutual presence of neediness results in alienation, separation, and the effort to eliminate need by domination of the roots of neediness. In a culture of criticism characterized by a hermeneutics of suspicion, this drive for domination is directed ostensibly at others, but more basically at oneself.[11]

The themes of presence and commitment are vitally important in a discussion of the community within a community that is formed by the relationship of doctor and patient. The physician is a member of a particular community, the profession of medicine, as well as a member of the community of everyday life that is shared with the patient and the family and all others. The two people may have no common center or commonly agreed upon purpose other than the one negotiated between them or the concern that brings them together. While there are various models of relationship that purport to exemplify the means to maximize the benefits of this community, the primary emphasis in a pluralistic world must lie on the assumptions of careful knowledge and trust between them. The present model of the medical relationship, grounded in the biomedical model of illness and enmeshed in a climate of suspicion, results in a relationship in which the giving and receiving of knowledge and trust are brokered by self-interested, self-seeking adversaries. Thus, the community of two is threatened.

Physicians do not care for the other because he or she carries the cross of illness. Hauerwas, in treating the theme of *imitatio*, writes that the cross is more than a "general symbol of the moral significance of self-sacrifice." Rather, "the cross is God's kingdom come" ([121], p. 87). If we set aside the notion that the cross represents God's selfless giving, and see it as symbolizing the violation of love, it is no longer the goal of love but an expression of the length to which God will go in order to restore broken community ([158], p. 20). The physician, then, cares for the patient because we all share a mutual neediness, represented by the commonality of illness; after all, the patient in the bed mirrors the physician's self. Physicians, therefore, make the commitment to be present in the face of, in spite of, pain and suffering. The continual presence of the other, and the corresponding demand for care and trust, cannot be denied or eliminated. Attempts to do so by idealizing selflessness and self-sufficiency demean the self and other, and lead to domination and self-annihilation. These attempts to absent otherness only create emptiness within the self that is experienced as self-hatred, inadequacy, and self-negation. Self-abnegation is not the central feature of fiduciary regard. By making it so, we distort the profession's tradition of service.

Therefore, a theological medical ethics must be appreciative of the actual experience of practitioners and attend more closely to the context in which physicians, nurses, patients, and others experience their moral lives. In other words, medical ethics must be attentive to the particularities of the practical setting, to the communities in which people dwell and from which they derive their values. Rather than focusing on the application of moral principles to health care issues, which are necessarily abstract and therefore not relevant to the particular circumstances of actual cases, another approach to an understanding of the important meaning of trust in physicians' "practice" focuses more on issues of community, interdependence, care, and the specific responsibilities of people with attention to the common good.

NOTES

[1] Ordinarily, the difference between the familiar and the unfamiliar is mediated by symbols that serve to bring the unfamiliar into the familiar while retaining something of its power. Luhman discusses the role of symbols in mediating the movement of the unfamiliar into the familiar [169].

[2] The biomedical model fosters such an effort to deny the existential condition of neediness inasmuch as it stresses "curing." We must accept, as Hauerwas argues, that medicine is a tragic profession since it deals with the conditions of life that cannot be eliminated [119].

[3] For a discussion of sin, see ([43], pp. 585–586). I am using the term in the Protestant sense of the actions and moral transgressions that result from a broken relationship with God in mistrust and a lack of faith (p. 585). The sin of pride does not refer to particular acts, but to the condition of estrangement that misleads humans in their exercise of freedom.

[4] Faith is not directed toward a God who acted or who will act; faith lies in the presence of God-as-Christ in the present. Barth, like Luther, sees Christian faith not primarily as belief in dogma or necessarily fixed and determinable events, but as wholehearted trust in divine grace and love as revealed in Jesus (see [196, 201]).

[5] See, for instance, a review of contract and covenant models of physician-patient interaction in [46, 184].

[6] The "we-relation" is used by Schutz to refer to the relation in which each person constitutes the other as an actor in the everyday world ([243], pp. 221–226, 287–356). It is through this "we-relationship" that people come to participate in reality. A reciprocal thou-orientation exists, a relationship of mutual, conscious awareness of the other ([228], pp. 171–172).

[7] The myth of origin, the expectation that explanation can be discovered, represents the attempt to efface human dependency by making loss and incompleteness "secondary" rather than "original" ([267], p. 71). The loss of independence becomes a "fall" from the original situation of complete Being (see [58]).

[8] "The Word, in order to touch me, must become warm flesh. Only then do I understand – when I can smell, see, and touch" ([150], p. 43).

[9] The vision of community presented by Buber is of a circle: "[The relations of persons] with their true 'Thou,' the radial lines that proceed from all points of the 'I' to the Center, form a circle. It is not the periphery, the community, that comes first, but the radii, the common quality of relation with the center. This alone guaranteed the authentic existence of the community" [61]. The radii indicate the relationship that must exist between each member of the community and the common center or purpose.

[10] "Buber's image of the circle is again appropriate, as the arcs that connect the persons along the periphery may be taken to represent the bonds or commitments between them" ([61], p. 192).

[11] For a discussion of the culture of criticism, see [110]. The expression "hermeneutics of suspicion" is attributed to the work of Karl Marx, Sigmund Freud, and Friedrich Nietzsche. The use of hermeneutics to "replenish" meaning is the intent of Paul Ricouer, Martin Heidegger, and Hans-Georg Gadamer ([110], p. 194). The role of biblical narrative in subverting and renewing culture is presented by [240].

CHAPTER 7

THE MUTUALLY SUSTAINING PRESENCE OF TRUST-AS-FAITH

Postmodernity emphasizes the dependence of all truth claims on the vested interest of the claimant, thus undermining any consensus of meaning [159]. We have focused on the challenge that the postmodern condition presents to the medical profession in an effort to formulate an orientation for trust-as-faith that can provide meaning in the shifting swirl of experience that comprises the physician-patient relationship. Any attempt to discover the original *idea* or essence of the profession is doomed to failure. Even the idea of such a truth is ambiguous in that it depends upon the social and intellectual conditions of the particular age ([192], pp. 222, 244–245). The answer to any question must depend on the way in which the question is asked, and there is no one way to ask questions of the past. The question will be formed from a subjective interpretation of available data, and the answers become momentary, relative.

Any attempt to conceptualize a definitive view of the essential traits or function that definitively defines the "good" physician implicitly becomes a criticism of the present conceptualization, and the criticism is advanced always from a particular point of view. Thus, it is important to understand the way in which the physician trusts his or her horizon of meaning to provide definitional limits to medical work. This horizon is constituted by confidence in medical knowledge and skill, and the taken-for-granted assumption by the physician that he or she is trustworthy. When this assumption is bracketed, however, the meaning of trust requires re-creation.

The medical profession has cultivated a degree of trustworthiness in the way its accent defines the practitioner's personal and professional self-image, and the "face" of competence and beneficence that it presents to the public. The way in which the physician regards his or her work is fiduciary in nature. While it enhances patient care, however, it also benefits the profession's prestige and authority and aids in monopolizing the health care market. In the last one hundred years, the fiduciary commitment of medicine has helped the medical profession organize itself socially and politically into an institution controlling its work, with the justification that this control is for the patient's good. The growth of a strong taken-for-granted attitude towards technical expertise (viewed as a "good" by the profession, as a major characteristic delineating the profession and distinguishing it from "quacks") turns the physician into a domineering figure who acts with the "best of intentions" while furthering his or her own interests and power.

Unfortunately, as a result of a distorted relationship that denies mutual dependency, a concern with the depth and commitment of physicians' fiducial nature is developing. I do not think that moral dilemmas are the result of this concern, but conflicts over autonomy and authority find expression in it and push the parties towards contract, regulation, and litigation as a means of resolving the issue. Legal language increases as a common understanding of trust is called into question.

Trust, as a response to others and a reliance on others, has something to do with relationships of mutual dependency ([166], p. 197). The medical professional is in a mutually dependent relationship with the patient and also with the public that "permits" or "licenses" his or her practice, and gives to him the trust that is fundamental to his work (literally, the trust that allows the doctor "to cut someone"). If mutual trust is to exist in social relations, each person must be willing to surrender something of his or her self-control and accept the vulnerability that accompanies the dialectical structure of social existence. This surrender goes beyond contractual confidence, yet stops short of blind faith. It leads to the nature of trust and faith as fides and fiducia in a covenantal relationship.

The covenantal relationship between doctor-patient that can create responsiveness and responsibility is a "tripartite concept: a covenant [which includes] not only an involvement with a partner in time, and a responsive contract, but the notion of a change in being. . . ." ([175], p. 69). Inasmuch as people are born into a world marked by an accent of reality developed over time, all that they know of the world is given to them from their social milieu. In this sense, biological life and a sense of "life" are both gifts. Receiving this gift is only one half of the relationship involved in the everyday world. Receiving is completed by the giving someone returns to the world. In contributing to society by our presence, thoughts, and actions, we add to the social matrix out of which society continues to develop. Just as we need the social realm in order to exist, we need to return something of ourselves to the world in order to exist meaningfully. "Receiving" becomes a correlate to "giving" in a relationship; the way we receive care is as important as the way in which we give it.[1] The three covenantal dimensions are included in the "we-relation," grounded in its constitutive elements of interdependent giving and receiving.

The natural attitude of the world into which people are born represents (or gives) a sense of reality to them through relationships of all kinds, principally social, and they in turn re-present (or give back) themselves to reality. Ideally, each participant is able to give and receive freely with hope for some equal degree of reciprocity. However, this exchange exists on any level: The person in a vegetative coma receives the gift of care on which his or her life may depend. While this person cannot give back an equal degree of animated response to the care-giver, nonetheless a relationship exists. The person who is in relationship with the patient-person does receive something (be it aggravation, financial burden, or unwanted responsibility). If the gift of the patient does not meet the needs of the caregivers, they may choose to give up their responsibility, although this decision is made in the context created by the "we-relationship." The attempt to design a system to weigh in some objective manner the benefits and burdens of any we-relationship is problematical and necessary, but if it becomes as mechanistic as the contractual relationship that grows out of a climate of distrust and self-protection, it will be inadequate.

The recognition of interdependence presents the possibility of a mutually "sustaining presence" in the medical relationship. In place of a univocal economy of domination, a dialogically oriented medical model reminds us that theology cannot look down upon human health and illness from some unaffected vantage point. Theology itself partly is dependent upon human experience for its structures of meaning. Every theological perception contains elements conditioned by human experience, and every human experience can be perceived and interpreted through some religious perspective. The difference between a univocal and a dialogic method is the difference between a theology of medical care in which the taken-for-granted perceptions of theology are imposed upon medicine, and a medical theology in which medicine suggests theological perspectives. There must be an appreciation of mutual presence in medical encounters.

However, this presence is not a passive experience.[2] Trust based on blind faith, on passive surrender to another, must be distinguished from trust that is earned after having first acknowledged to oneself, and then shared with the other, what one knows and does not know about the decision to be made in an uncertain situation. Interdependent human relationship is reciprocal and mutually enlivening ([28], pp. 58, 61, 67). In the we-relationship, encounter is juxtaposed to experience. For Buber, the "It" is experienced, while the "Thou" is encountered. The Thou cannot be "experienced" as if an objective event. For Buber the Thou is partly an It, but also is much more. In encounter there is no mediator between the Thou and the I. The two are necessary for the completion of the other ([28], p. 62). The encounter is a present event, and for as long as it lasts, the present is actual and complete. One form confronts and fulfills the other in a "we-relation."

As we have seen, the world of daily life is made up of multiple levels of reality or provinces of meaning. Each of the finite provinces of meaning may receive the accent of reality, may be attended to as real. There is no formula of transformation that enables one to pass smoothly from one province to another. The transition is always experienced as a shock as one moves from one accent with its own concerns, language, objects of relevance, actions, and habits to another one. Only by a "leap" is that passage possible ([243], pp. 232–259). Giving up the meaning structures of one province is an unsettling experience. The perception of choices and the freedom to choose involve a person in existential anxiety. It is tempting to escape this anxiety by renouncing the struggle for self-assertion by seeking self-security and domination.

In the face of the uncertainty of dependency and vulnerability, the "will to accept responsibility," the will to respond to others, requires courage and trust. Trust is not the blind acceptance of risk; it is the courageous choice to rely on another in spite of the threat of dependency. Earl Shelp writes of an "alliance of uncertainty" that creates a bond between physicians and patients, promoting a mutual compassion that makes presence possible to each other [250]. The leap of faith that enables a person to be present to another represents a moral effort to trust that includes the "courage to fail," the courage to be vulnerable and dependent on others [78].[3]

Superficially, the development of a technologically or scientifically based health care is motivated by the desire to satisfy our desires (wants) in regard to health. On a deeper level, though, the development is fueled by the desire to eliminate our dependency. The shift in self-conception by the profession (and the public's conception of the profession) from that of an art to that of a science fosters the development of domination. Research into disease, the growth of medical technology and technique, contributes to and fuels the view that the physician is the life-giver or denier of death. This view gives the medical professional a great deal of power within the natural attitude of the everyday world. However, this gift of power is shaped by, and in turn shapes, the view of the relationship between life and death. If it is fueled by the denial of dependency, it can become domineering and oppressive. The doctor who is seen as the giver of life struggles in the face of death to meet the needs of the patient through the use of technology and the manipulation of information. The patient fearing death who gives away control, autonomy, and responsibility is unable to receive death meaningfully if it should appear. A paternalistic relationship can be the result of a denial of neediness in both parties, and worse, the physician may make himself absent from the patient out of a sense of shame at having failed to deny death (itself a denial contained within the attitude of a profession that has denied its own neediness in the face of the correlative gift of life and death). Trust is possible only in the acknowledgment of suffering. Any attempt to rid the world of suffering is Promethean ([258], p. 275).

Dependency cannot be eliminated. The concern for human dignity, self control, and autonomy all represent the inability to accept and express mutual needs. The turn to consumerism as the basis for the professional-client relationship may counter philanthropic conceit, but it also may be the denial of neediness. A need for communication and relationship can be replaced by the demand for accountability in the contractual arrangement. The only needs covered are those specified within the contract. The physician becomes a caretaker held accountable by the one who pays the bills. Within such an arrangement there does not seem to be the possibility for the nurturance of body and soul called for by the covenantal relationship. However, physician and patient are correlative entities; each gives and receives purpose and meaning from the other. They can become a "sustaining presence" to each other.

A medical theological ethics holds out the "faith" that medicine can be responsive and altruistic, without becoming paternalistic and domineering. The physician who is able to give up the need for power by accepting the ambiguities and limitations of his or her "art" is better able to accept the needs and concerns of patients. Since every act of giving implies a loss, to be able to accept one's own neediness and incompleteness enables one to give more selflessly, not out of the conceit of denial, but out of a sense of mutual concern and caring. The person who acknowledges his or her dependency is better able to face the responsibility and obligations of the patient, who must relinquish some self-control to the doctor in trusting acceptance of the caring relationship. In the covenantal model, the doctor who gives care receives trust; the patient who gives trust receives care. Each one finds some ontological fulfillment in the needs of the other regardless of the outcome of treatment. A shared view, a mutual presence, replaces the single view of the expert in a relationship of "fidelity that exceeds any specification" ([175], p. 70).

NOTES

[1] Keith-Lucas recognizes that an important characteristic of the helping relationship is that it is mutual, not merely one-way ([153], p. 47). The doctor gives of his or her expertise and dedication, yet receives in return from the patient, among other things, the "license" to practice the art.

[2] Hauerwas uses the term "suffering presence" to describe the traditional dedication of physicians to being present to those in need, even when nothing can be done (see [123]). Shelp develops the phrase "sustaining presence" as a corrective to the image of the physician as an all powerful, authoritarian figure. The image of the doctor as "sustainer" allows for the recognition of the profession's superior knowledge and expertise while emphasizing the fiducial commitment to remain in relationship with patients even when nothing can be done (see [252]). Both authors reject the abandonment of the patient by the physician who finds that the possibilities of active treatment are gone. Both see presence as an active condition, enlivening the relationship between both parties even in a situation where intervention is useless.

[3] For a discussion of Kierkegaard's idea of the "leap of faith," see [48, 259].

REFERENCES

1. Abrams, F.R.: 1986, "Caring for the Sick: An Emerging Industrial By-Product", *Journal of the American Medical Association* 255, 937–938.
2. Altizer, T.J.J.: 1987, *The Self-Embodiment of God*, University Press of America, Lanham, MD.
3. Ashley, B.M., and O'Rourke, K.D.: 1978, *Health Care Ethics*, The Catholic Hospital Association, St. Louis, MO.
4. Atkinson, P.: 1981, *The Clinical Experience: The Construction and Reconstruction of Medical Reality*, Gower, Farnborough, England.
5. Atkinson, P.: 1983, "The Reproduction of the Professional Community", in R. Dingwall and P. Lewis (eds.), *The Sociology of the Professions*, The Macmillan Press, London, pp. 224–241.
6. Bagdanich, W., and Waldholz, M.: 1989, "Patients for Sale", *The Wall Street Journal* 83 (February 27), A1, A4.
7. Barber, B.: 1963, "Some Problems in the Sociology of Professions", *Daedelus* 92, 669–688.
8. Barber, B.: 1983, *The Logic and Limits of Trust*, Rutgers University Press, Rutgers, New Jersey.
9. Barber, M.: 1986, "Alfred Schutz's Methodology and the Paradox of the Sociology of Knowledge", *Philosophy Today* 30, 58–65.
10. Baron, R.J.: 1985, "An Introduction to Medical Phenomenology: I Can't Hear You While I'm Listening", *Annals of Internal Medicine* 103, 606–611.
11. Baron, R.J.: 1989, "Dogmatics, Empirics, and Moral Medicine", *Hastings Center Report* 19, 41–42.
12. Barth, K.: 1975, *Church Dogmatics*, Clark, Edinburgh.
13. Beauchamp, T., and Walters, L.: 1982, *Contemporary Issues in Bioethics*, Wadsworth, Belmont, CA.
14. Becker, H.K. *et al.*: 1961, *The Boys in White*, University of Chicago Press, Chicago.
15. Bellah, R. *et al.*: 1985, *Habits of the Heart: Individualism and Commitment in American Life*, University of California Press, Berkeley, CA.
16. Bennett, W.: 1980, "Getting Ethics", *Commentary* (December), 62–65.
17. Berlant, J.: 1975, *Profession and Monopoly*, University of California Press, Berkeley.
18. Berlant, J.: 1977, "Medical Ethics and Monopolization", in S.J. Reiser *et al.* (eds.), *The Ethics of Medicine*, MIT Press, Cambridge, MA., pp. 52–64.
19. Blendon, R.J.: 1988, "The Public's View of the Future of Health Care", *Journal of the American Medical Association* 259, 3587–3593.
20. Bok, S.: 1978, *Lying: Moral Choice in Public and Private Life*, Pantheon Books, New York.
21. Bosk, C.L.: 1992, *All God's Children: Genetic Counseling in a Pediatrics Hospital*, University of Chicago Press, Chicago.
22. Brand, P., and Yancey, P.: 1993, *Pain: The Gift Nobody Wants*, Zondervan, New York.
23. Brandt, A.M.: 1991, "Emerging Themes in the History of Medicine", *The Milbank Quarterly* 69(2), 199–214.
24. Brenneman, W., Jr. *et al.*: 1982, *The Seeing Eye: Hermeneutical Phenomenology in the Study of Religion*, The Pennsylvania State University Press, University Park, PA.
25. Bricker, E.M.: 1989, "Industrial Marketing and Medical Ethics", *New England Journal of Medicine* 320 (June 22), 1680–1692.

26. Brown, N.: 1959, *Life Against Death: The Psychoanalytic Meaning of History*, Random House, New York.
27. Brunner, E.: 1937, *The Divine Imperative*, The Westminster Press, Philadelphia.
28. Buber, M.: 1958, *I and Thou*, Charles Scribner's Sons: New York.
29. Bucher, R.: 1970, "Social Process and Power in Medical School", in M. Zald (ed.), *Power in Organizations*, Vanderbilt University Press, Nashville.
30. Bucher, R., and Strauss, A.: 1961, "Professions in Process", *American Journal of Sociology* 66 (January), 325–334.
31. Bullough, V.: 1972, "A Brief History of Medical Practice", in E. Freidson and J. Lorber (eds.), *Medical Men and Their Work*, Aldine-Atherton, Chicago, pp. 86–102.
32. Cahoone, L.E.: 1988, *The Dilemma of Modernity: Philosophy, Culture, and Anti-culture*, State University of New York Press, New York.
33. Camenisch, P.F.: 1981, "Gift and Gratitude in Ethics", *Journal of Religious Education* 9, 1–34.
34. Campbell, A.V.: 1984a, *Moderated Love – A Theology of Professional Care*, SPCK, London.
35. Campbell, A.V.: 1984b, *Professional Care: Its Meaning and Practice*, Fortress Press, Philadelphia.
36. Camporesi, P.: 1989, "The [illegible] Host: A Wondrous Excess", in M. Feher (ed.), *Fragments of a History of the Human Body*, The MIT Press, Cambridge, MA.
37. Carlton, W.: 1978, *"In Our Professional Opinion . . ." – The Primacy of Clinical Judgment Over Moral Choice*, University of Notre Dame Press, Notre Dame.
38. Carmody, J.: 1993, *How to Handle Trouble: A Guide to Peace of Mind*, Doubleday, New York.
39. Casberg, M.A.: 1968, "The Effect of Specialization on the Treatment of the Whole Man", in Dale White (ed.), *Dialogue in Medicine and Theology*, Abingdon Press, New York, pp. 81–96.
40. Cassell, E.J.: 1985, *The Healer's Art*, MIT Press, Cambridge, MA.
41. Childress, J.: 1981, *Priorities in Biomedical Ethics*, The Westminster Press, Philadelphia.
42. Childress, J.: 1986a, "Shame", in J.F. Childress and J. MacQuarrie (eds.), *The Westminster Dictionary of Christian Ethics*, The Westminster Press, Philadelphia, pp. 583–584.
43. Childress, J.: 1986b, "Trust", in J.F. Childress and J. MacQuarrie (eds.), *The Westminster Dictionary of Christian Ethics*, The Westminster Press, Philadelphia, pp. 632–633.
44. Christiansen, D.: 1980, "The Elderly and Their Families: The Problem of Dependence", *New Catholic World* 223, 100–104.
45. Churchill, L.R.: 1989, "Trust, Autonomy, and Advanced Directives", *Journal of Religion and Health* 28, 175–183.
46. Clouser, K.D.: 1983, "Veatch, May, and Models", in E.E. Shelp (ed.), *The Clinical Encounter*, Kluwer Academic Publishers, Dordrecht, Holland, pp. 89–104.
47. Cochrane, C.: 1957, *Christianity and Classical Culture*, Galaxy, New York.
48. Collins, J.: 1962, "Faith and Reflection in Kierkegaard", in H.A. Johnson and N. Thulstrop (eds.), *A Kierkegaard Critique*, Henry Regnery Company, Chicago, pp. 141–155.
49. Colombotos, J., and Kirchner, C.: 1986, *Physicians and Social Change*, Oxford University Press, New York.
50. Coombs, R.H.: 1978, *Mastering Medicine: Professional Socialization in Medical School*, The Free Press, New York.
51. Cotton, D.J.: 1988, "The Impact of AIDS on the Medical Care System", *Journal of the American Medical Association* 260 (July 22/29), 519–523.
52. Dan, B.: 1987, "Patients Without Physicians: The New Risk of AIDS", *Journal of the American Medical Association* 258, 1940.
53. Daniel, S.L.: 1986, "The Patient as Text: A Model of Clinical Hermeneutics", *Theoretical Medicine* 7, 195–210.
54. Daniels, A.K.: 1973, "How Free Should Professions Be?", in E. Freidson (ed.), *The Professions and Their Prospects*, Sage, Beverly Hills, CA., pp. 39–57.

55. Dasgupta, P.: 1988, "Trust as Commodity", in D. Gambetta (ed.), *Trust: Making and Breaking Cooperative Relations*, Basil Blackwell, London, pp. 51–52.
56. Derrida, J.: 1976, *Of Grammatology*, trans. G.C. Spivak, Johns Hopkins University Press, Baltimore.
57. Derrida, J.: 1978, *Writing and Difference*, trans. A. Bass, University of Chicago Press, Chicago.
58. DiCenso, J.J.: 1988, "Heidegger's Hermeneutics of Fallenness", *Journal of the American Academy of Religion* 64, 667–679.
59. Donoghue, D.: 1988, "Insincerity and Authenticity", *The New Republic* 198 (February 15), 26.
60. Dostoyevsky, F.: 1976, *The Brothers Karamazov*, trans. Constance Garnett, W.W. Norton, New York.
61. Duffy, M.F.: 1988, "The Challenge to the Christian Community", *Religious Education* 83, 191–199.
62. Dunn, J.: 1988, "Trust and Political Agency", in D. Gambetta (ed.), *Trust: Making and Breaking Cooperative Relations*, Basil Blackwell, London, pp. 73–93.
63. Dunn, M.R.: 1992, *American Medical News* (June 1), 28.
64. Dyer, A.: 1985, "Virtue and Medicine: A Physician's Analysis", in E.E. Shelp (ed.), *Virtue and Medicine*, D. Reidel Publishing Company, Dordrecht, pp. 223–235.
65. Eco, U.: 1985, "Strategies of Lying", in M. Blonsky (ed.), *On Signs*, The John Hopkins University Press, Baltimore, MD., pp. 3–11.
66. Eisenberg, L.: 1977, "The Search for Care", *Daedalus* 106, 235–246.
67. Elliot, P.: 1972, *The Sociology of the Professions*, Macmillan, London.
68. Engel, G.L.: 1977, "The Need for a New Medical Model: A Challenge for Biomedicine", *Science* 196, 129–136.
69. Engelhardt, H.T., Jr.: 1982b, "Rights and Responsibilities of Patients and Physicians", in Beauchamp, T.L., and Walters, L.: *Contemporary Issues in Bioethics*, Wadsworth, Belmont, CA.
70. Engelhardt, H.T., Jr.: 1986, *The Foundations of Bioethics*, Oxford University Press, New York.
71. Engelhardt, H.T., Jr., and Rie, M.A.: 1988, "Morality for the Medical-Industrial Complex", *New England Journal of Medicine* 319, 1086–1089.
72. Feuerbach, L.: 1957, *The Essence of Christianity*, Harper and Row, New York.
73. Fisher, S.: 1986, *In the Patient's Best Interest: Women and the Politics of Medical Decisions*, Rutgers University Press, New Jersey.
74. Foss, L.: 1989, "The Challenge to Biomedicine: A Foundations Perspective", *Journal of Medicine and Philosophy* 14, 165–191.
75. Foucault, M.: 1973, *The Birth of the Clinic: An Archeology of Medical Perception*, Pantheon, New York.
76. Fox, R.: 1957, "Training for Uncertainty", in R. Merton *et al.*, *The Student Physician*, Harvard University Press, Cambridge, pp. 207–241.
77. Fox, R.: 1979, *Essays in Medical Sociology*, John Riley and Sons, New York.
78. Fox, R., and Swazey, J.: 1974, *The Courage to Fail*, University of Chicago Press, Chicago.
79. Frank, A.W.: 1991a, "From Sick Role to Health Role: Deconstructing Parsons", in R. Robertson and B.S. Turner (eds.), *Talcott Parsons: Theorist of Modernity*, Sage, London, pp. 205–216.
80. Frank, A.W.: 1991b, *At the Will of the Body*, Houghton Mifflin, Boston.
81. Frank, A.W.: 1994, "Book Review", *Second Opinion* 19(3), 102–107.
82. Freidson, E.: 1970a, *Profession of Medicine: A Study of the Sociology of Applied Knowledge*, Dodd, Mead, New York.
83. Freidson, E.: 1970b, *Professional Dominance: The Social Structure of Medical Care*, Aldine-Atherton Press, New York.
84. Freidson, E.: 1973, *The Professions and Their Prospects*, Sage, Beverly Hills, CA.

85. Freidson, E.: 1975, *Doctoring Together: A Study of Professional Social Control*, Elsevier, New York.
86. Freidson, E.: 1977, "The Future of Professionalization", in M. Stacey *et al.* (eds.), *Health and the Division of Labor*, Croon Helm, London, pp. 14–28.
87. Freidson, E.: 1986, *Professional Powers: A Study of the Institutionalization of Formal Knowledge*, University of Chicago Press, Chicago.
88. Freire, P.: 1971, *A Pedagogy of the Oppressed*, Herder and Herder, New York.
89. Freshnock, L.: 1984, *Physicians and Public Attitudes on Health Care Issues*, The American Medical Association, Chicago.
90. Freyman, J.G.: 1974, *The American Health Care System: Its Genesis and Trajectory*, Medcom, New York.
91. Gadow, S.: 1982, "Body and Self: A Dialectic", in V. Kestenbaum (ed.), *The Humanity of the Ill*, University of Tennessee Press, Knoxville, pp. 86–100.
92. Gambetta, D.: 1988, "Can We Trust Trust?", in D. Gambetta (ed.), *Trust: Making and Breaking Cooperative Relations*, Basil Blackwell, London, pp. 213–237.
93. Garfinkel, H.: 1967, *Studies in Ethnomethodology*, Prentice-Hall, Englewood Cliffs, New Jersey.
94. Gay, P.: 1977, *The Enlightenment: An Interpretation*, Vol. 1, Norton, New York
95. Gilkey, L.: 1985, "Theological Frontiers: Implications for Bioethics", in E.E. Shelp (ed.), *Theology and Bioethics*, D. Reidel Publishing Company, Dordrecht, Holland, pp. 115–134.
96. Gillon, R.: 1986, "More on Professional Ethics", *Journal of Medical Ethics* 12, 2, 59–60.
97. Goetz, R.: 1989, "On Blasphemy: Advice for the Ayatollah", *The Christian Century* (March 8), 253–255.
98. Golden, J., and Johnston, G.: 1979, "Problems of Distortion in Doctor-Patient Communications", *Psychiatry in Medicine* 1, 127–149.
99. Good, D.: 1988, "Individuals, Interpersonal Relations, and Trust", in D. Gambetta (ed.), *Trust: Making and Breaking Cooperative Relations*, Basil Blackwell, London, pp. 31–48.
100. Goode, W.J.: 1957, "Community Within a Community: The Professions", *American Sociological Review* 22, 194–200.
101. Gordy, M.: 1978, "Sociality", in W. Reich (ed.), *Encyclopedia of Bioethics* 4, The Free Press, New York, pp. 1603–1606.
102. Gorman, J.: 1992, "Take A Little Deadly Nightshade: and You'll Feel Better", *New York Times Magazine* (August 30), Section 6, 23–28, 73.
103. Gorman, R.: 1977, *The Dual Vision: Alfred Schutz and the Myth of Phenomenological Social Science*, Routledge & Kegan Paul, London.
104. Gorovitz, S.: 1986, "Baiting Bioethics", *Ethics* 96, 356–374.
105. Gorovitz, S., and MacIntyre, A.: 1976, "Toward a Theory of Medical Fallibility", in H.T. Engelhardt, Jr., and D. Callahan (eds.), *Science, Ethics, and Medicine* I, Hastings Center Publication, Hastings-on-Hudson, NY., pp. 248–274.
106. Gould, S.J.: 1981, *The Mismeasure of Man*, W.W. Norton, New York.
107. Greenberg, D.: 1974, "Ethics and Nonsense", *New England Journal of Medicine* 290, 977–978.
108. Griffin, D.R., 1980, "The Holy, Necessary Goodness, and Morality", *Journal of Religious Ethics* 8 (Fall), 330–349.
109. Guarino, T.: 1993, "Between Foundationalism and Nihilism: Is *Phronēsis* the *Via Media* for Theology?", *Theological Studies* 54, 37–54.
110. Gunn, G.: 1987, *The Culture of Criticism and the Criticism of Culture*, Oxford University Press, New York.
111. Gurvitch, G.: 1971, *The Social Framework of Knowledge*, Harper and Row, New York.
112. Gustafson, J.: 1975, *Can Ethics be Christian?*, University of Chicago Press, Chicago.
113. Gustafson, J.: 1984, *Ethics From a Theocentric Perspective*, Vol 2, University of Chicago Press, Chicago.
114. Harnack, A.: 1978, *What is Christianity?*, Peter Smith, Gloucester, MA.
115. Harned, D.B.: 1973, *Faith and Virtue*, Pilgrim Press, Philadelphia.

116. Hartshorne, C., and Reese, W.L.: 1953, *Philosophers Speak of God*, University of Chicago Press, Chicago.
117. Hartt, J.N.: 1986, "Faith", in J.F. Childress and J. MacQuarrie (eds.), *The Westminster Dictionary of Christian Ethics*, The Westminster Press, Philadelphia, pp. 222–224.
118. Hatlie, M.: 1989, "Professional Liability", *Journal of the American Medical Association* 261: 2881–2882.
119. Hauerwas, S.: 1977, *Truthfulness and Tragedy*, University of Notre Dame Press, Notre Dame.
120. Hauerwas, S.: 1981, *A Community of Character: Toward a Constructive Christian Social Ethic*, University of Notre Dame Press, Notre Dame.
121. Hauerwas, S.: 1983, *The Peaceable Kingdom: A Primer in Christian Ethics*, University of Notre Dame Press, Notre Dame.
122. Hauerwas, S.: 1985, "Salvation and Health: Why Medicine Needs the Church", in E.E. Shelp (ed.), *Theology and Bioethics*, D. Reidel Publishing Company, Dordrecht, Holland, pp. 205–224.
123. Hauerwas, S.: 1986, *Suffering Presence: Theological Reflections on Medicine, the Mentally Ill, and the Church*, University of Notre Dame Press, Notre Dame.
124. Haug, M., and Sussman, M.: 1969, "Professional Autonomy and the Revolt of the Client", *Social Problems* 17, 153–161.
125. Hawthorn, G.: 1988, "Three Ironies in Trust," in D. Gambetta (ed.), *Trust: Making and Breaking Cooperative Relations*, Basil Blackwell, London, pp. 111–112.
126. Hays, R.D., and Ware, J.E.: 1986, "My Medical Care is Better Than Yours", *Medical Care* 24, 519–524.
127. Heidegger, M.: 1962, *Being and Time*, trans. J. Macquarrie and E. Robinson, Harper and Row, New York.
128. Henslin, J.: 1972, "What Makes for Trust?", in J. Henslin (ed.), *Down to Earth Sociology*, Free Press, New York, pp. 1–22.
129. Herman, S.W.: 1992, "The Modern Business Corporation and an Ethics of Trust", *The Journal of Religious Ethics* 20 (Spring), 111–148.
130. Hilfiker, D.: 1985, *Healing the Wounds: A Physician Looks at His Work*, Pantheon Books, New York.
131. Holleman, W.L.: 1992, "Challenges Facing Student-Clinicians", *Humane Medicine* 8, 205–211.
132. Holler, L.: 1989, "Is There a Thou 'Within' Nature? A Feminist Dialogue with H. Richard Niebuhr", *The Journal of Religious Ethics* 17, 81–102.
133. Holm, S.: 1988, "The Peaceable Pluralistic Society and the Question of Persons", *Journal of Philosophy and Medicine* 13, 379–386.
134. Horobin, G.: 1983, "Professional Mystery: The Maintenance of Charisma in General Medical Practice", in R. Dingwall and P. Lewis (eds.), *The Sociology of the Professions*, The Macmillan Press, London, pp. 84–105.
135. Hulka, B.S. *et al.*: 1975, "Correlates of Satisfaction and Dissatisfaction with Medical Care: A Common Perspective", *Medical Care* 13, 648–658.
136. Husserl, E.: 1960, *Cartesian Meditations*, trans. D. Cairns, Martinus Nijhoff, The Hague.
137. Husserl, E.: 1962, *Ideas: General Introduction to Pure Phenomenology*, Collier Books, New York.
138. Husserl, E.: 1964, *The Idea of Phenomenology*, trans. W. Alston and G. Nakhnikian, Martinus Nijhoff, The Hague.
139. Husserl, E.: 1970, *The Crisis of European Sciences and Transcendental Phenomenology*, trans. David Carr, Northwestern University Press, Evanston, Ill.
140. Hyman, D.A., and Williamson, J.V.: 1989, "Fraud and Abuse: Setting the Limits on Physicians' Entrepreneurship", *New England Journal of Medicine* 320, 1275–1278.
141. Illich, I.: 1976, *Medical Nemesis*, Pantheon, New York.
142. Ingelfinger, F.J.: 1976, "The Physician's Contribution to the Health Care System", *New England Journal of Medicine* 295, 565.

143. James, W.: 1890, *Principles of Psychology* 2, Henry Holt Company, New York.
144. Jewson, N.D.: 1974, "The Disappearance of the Sick Man From the Medical Cosmology", *Sociology* 8, 369–385.
145. Johnson, T.J.: 1972, *Professions and Power*, Macmillan, New York.
146. Jonas, H.: 1966, *The Phenomenology of Life: Towards a Philosophical Biology*, Harper Row, New York.
147. Karen, R.: 1992, "Shame," *The Atlantic Monthly* 290, 40ff.
148. Kass, L.: 1983, "Professing Ethically: On the Place of Ethics in Defining Medicine", *Journal of the American Medical Association* 249, 1306–1307.
149. Katz, J.: 1984, *The Silent World of Doctor and Patient*, Macmillan, Free Press, New York.
150. Kazantzakis, N.: 1965, *Report to Greco*, Simon and Schuster, New York.
151. Kaufman, G.: 1975, *An Essay on Theological Method*, Scholars Press, Missoula, MO.
152. Keith, J.N.: 1991, "Healing in a Theological Perspective", in E.E. Shelp and R.H. Sunderland (eds.), *The Pastor as Counselor*, The Pilgrim Press, New York.
153. Keith-Lucas, A.: 1972, *Giving and Taking Help*, The University of North Carolina Press, Chapel Hill.
154. Kelber, W.: 1983, *The Oral and the Written Gospel*, The Fortress Press, Philadelphia.
155. Kermode, F.: 1979, *The Genesis of Secrecy: On the Interpretation of Narrative*, Harvard University Press, Cambridge, MA.
156. Kestenbaum, V.: 1982, *The Humanity of the Ill*, The University of Tennessee Press, Knoxville, TN.
157. Kierkegaard, S.: 1941, *Concluding Unscientific Postscript*, trans. D.F. Swenson and W. Lowrie, Princeton University Press, Princeton.
158. King, M.L.: 1986, *A Testament of Hope: The Essential Writings of Martin Luther King, Jr.*, J.M. Washington (ed.), Harper and Row, San Francisco.
159. Klemm, D.: 1986, *The Interpretation of Existence*, Scholars Press, Atlanta.
160. Kuhn, T.: 1970, *The Structure of Scientific Revolutions*, University of Chicago Press, Chicago.
161. Lacsamana, R.G.: 1985, "Cracks in the Mirror or What Patients Think of Us", *Journal of the Florida Medical Association* 72, 573–575.
162. Lash, A.: 1986, "Public Attitudes Towards Physicians", *Indiana Medical Journal* 79, 184–186.
163. Lazare, A.: 1987, "Shame and Humiliation in the Medical Encounter", *Archives of Internal Medicine* 147, 1653–1658.
164. Levin, D.M.: 1985, *The Body's Recollection of Being*, Routledge & Kegan Paul, Boston.
165. Loewy, E.: 1989, "Beneficence in Trust", *Hastings Center Report* 19 (January/February), 42–43.
166. Lorenz, E.H.: 1988, "Neither Friends Nor Strangers", in D. Gambetta (ed.), *Trust: Making and Breaking Cooperative Relations*, Basil Blackwell, London, pp. 194–210.
167. Luckman, T.: 1973, "Philosophy, Science, and Everyday Life", in M. Natanson (ed.), *Phenomenology and the Social Sciences*, Northwestern University Press, Evanston, Ill., pp. 143–185.
168. Luhman, N.: 1979, *Trust and Power*, Wiley, New York.
169. Luhman, N.: 1988, "Familiarity, Confidence, and Trust: Problems and Alternatives", in D. Gambetta (ed.), *Trust: Making and Breaking Cooperative Relations*, Basil Blackwell, London, pp. 94–108.
170. Lundberg, G.: 1985, "Medicine – A Profession In Trouble?", *Journal of the American Medical Association* 253, 2879–2880.
171. Lyotard, J.: 1984, *The Postmodern Condition: A Report on Knowledge*, University of Minneapolis Press, Minneapolis.
172. Macquarrie, J.: 1977, *Principles of Christian Theology*, Charles Scribner's Sons, New York.
173. Marzuk, P.: 1985, "The Right Kind of Paternalism", *New England Journal of Medicine* 313, 1474–1476.

174. May, W.F.: 1975, "Code and Covenant or Philanthropy and Contract?", *Hastings Center Report* 5: 29–38.
175. May, W.F.: 1977, "Code and Covenant or Philanthropy and Contract?", in S.J. Reiser *et al.* (eds.), *Ethics in Medicine*, MIT Press, Cambridge, MA., pp. 65–76.
176. May, W.F.: 1983, *The Physician's Covenant: Images of the Healer in Medical Ethics*, Westminster Press, Philadelphia.
177. Mazur, A.: 1973, "Disputes Between Experts", *Minerva* 11, 243–262.
178. McCormick, R.A.: 1989, "Theology and Bioethics", *Hastings Center Report* 19 (March/April), 5–10.
179. McCormick, R.A.: 1991, "Physician-assisted Suicide: Flight from Compassion", *The Christian Century* 108 (December), 1132–1134.
180. McCormick, R.A.: 1994, "Beyond Principlism Is Not Enough: A Theologian Reflects on the Real Challenge for U.S. Biomedical Ethics", in E. DuBose, R. Hamel, and L. O'Connell (eds.), *A Matter of Principles?*, Trinity Press International, Philadelphia, pp. 348–365.
181. McCullough, L.B.: 1983, "Modern Anglo-American Medical Ethics", in E.E. Shelp (ed.), *The Clinical Encounter*, D. Reidel Publishing Company, Dordrecht, Holland, pp. 56–57.
182. McGill, A.: 1982, *Suffering: A Test of Theological Method*, The Westminster Press, Philadelphia.
183. McIntyre, A.: 1981, *After Virtue*, University of Notre Dame Press, Notre Dame.
184. McKinney, G.P., and Sande, J.R.: 1993, *Theological Analyses of the Clinical Encounter*, Kluwer Academic Publishers, Dordrecht.
185. Mechanic, D.: 1985, "Public Perceptions of Medicine", *New England Journal of Medicine* 312, 181–183.
186. Mehan, H., and Wood, H.: 1971, *The Reality of Ethnomethodology*, Wiley, New York.
187. Merton, R.K. *et al.*: 1957, *The Student-Physician: Introductory Studies in the Sociology of Medical Education*, Harvard University Press, Cambridge, MA.
188. Merton, R.K., and Barber, E., 1963, "Sociological Ambivalence", in E. Tiryakin (ed.), *Sociological Theory, Values and Sociocultural Change: Essays in Honor of Pitirim Sorokin*, The Free Press, New York, pp. 91–120.
189. Miller, S.J.: 1970, *Prescription for Leadership: Training for the Medical Elite*, Aldine, Chicago.
190. Mishler, E.: 1981, "Viewpoint: Critical Perspectives on the Biomedical Model", in E. Mishler (ed.), *Social Contexts of Health, Illness, and Patient Care*, Cambridge University Press, Cambridge, England, pp. 1–23.
191. Moore, W.E., and Rosenblum, G.W.: 1970, T*he Professions: Roles and Rules*, Russell Sage Foundation, New York.
192. Morgan, R., and Pye, M.: 1977, *Ernst Troeltsch: Writings on Theology and Religion*, John Knox Press, Atlanta.
193. Morison, R.S.: 1981, "Bioethics After Two Decades", *Hastings Center Report* 11, 8–12.
194. Morreim, E.H.: 1988, "Cost Containment: Challenging Fidelity and Justice", *Hastings Center Report* 18, 20–25.
195. Morris, H.: 1971, *Guilt and Shame*, Wadsworth Publishing Company, Belmont, CA.
196. Mouw, R.J.: 1979, "Biblical Revelation and Medical Decisions", *Journal of Medicine and Philosophy* 4, 367–382.
197. Natanson, M.: 1962, "Introduction", in Alfred Schutz, *Collected Papers* 1, Martinus Nijhoff, The Hague.
198. Natanson, M.: 1968, *Literature, Philosophy, and the Social Sciences: Essays in Existentialism and Phenomenology*, Martinus Nijhoff, The Hague.
199. Natanson, M.: 1973, "Phenomenology and the Social Sciences", in M. Natanson (ed.), *Phenomenology and the Social Sciences*, Northwestern University Press, Evanston, IL., pp. 3–44.
200. Nelson, J.: 1987, "Reuniting Sexuality and Spirituality", *The Christian Century* (Feb. 25, 1987), 187–190.
201. Niebuhr, H.R.: 1951, *Christ and Culture*, Harper Torchbooks, New York.

202. Niebuhr, H.R.: 1960, *Radical Monotheism and Western Culture*, Harper and Row, New York.
203. Niebuhr, R.: 1947, *The Children of Light and The Children of Darkness*, Charles Scribner's Sons, New York.
204. Niebuhr, R.: 1964, *The Nature and Destiny of Man*, C. Scribner's Sons, New York.
205. Nietzsche, F.: 1961, *The Will to Power*, W. Kaufman (ed.), trans. W. Kaufman and R.J. Hollingdale, Vintage, New York.
206. Nietzsche, F.: 1966, *Beyond Good and Evil*, trans. W. Kaufman, Vintage, New York.
207. Nietzsche, F.: 1969, *On the Genealogy of Morals*, trans. W. Kaufman, Random House, New York.
208. Nisbet, R.: 1979, *History of the Idea of Progress*, Basic Books, New York.
209. Ogletree, T.: 1985, *Hospitality to the Stranger*, Fortress Press, Philadelphia.
210. Ottati, D.: 1983, "Reconstructing Christian Theology", *Religious Studies Review* 9 (July), 222–227.
211. Page, L.: 1992, "A Hard Look in the Mirror", *AMA News* 35(2), 37.
212. Paget, M.A.: 1993, *A Complex Sorrow: Reflections on Cancer and an Abbreviated Life*, M.L. DeVault (ed.), Temple University Press, Philadelphia.
213. Palmer, R.E.: 1969, *Hermeneutics: Interpretation Theory in Schleiermacher, Dilthey, Heidegger, and Gadamer*, Northwestern University Press, Evanston.
214. Parsons, T.: 1939, "The Professions and Social Structure", *Social Forces* 17, 457–467.
215. Parsons, T.: 1951a, *Essays in Sociological Theory*, The Free Press, Glencoe, IL.
216. Parsons, T.: 1951b, *The Social System*, The Free Press, Glencoe, IL.
217. Parsons, T.: 1968, "Professions", in D. Sills (ed.), *The International Encyclopedia of the Social Sciences*, Vol. 12, Macmillan and Free Press, New York, pp. 536–547.
218. Parsons, T.: 1969a, *Politics and Social Structure*, Free Press, New York.
219. Parsons, T.: 1969b, "Research with Human Subjects and the 'Professional Complex'", *Daedalus* 98 (Spring), 325–360.
220. Parsons, T.: 1975, "The Sick Role and the Role of the Physician Reconsidered", *Milbank Memorial Fund Quarterly* (Summer), 257–278.
221. Pellegrino, E.: 1974, "Medicine and Philosophy: Some Notes on the Flirtations of Minerva and Aesculapius", reprint, Society for Health and Human Values, Philadelphia. PA.
222. Pellegrino, E.: 1982, "Being Ill and Being Healed," in V. Kestenbaum (ed.), *The Humanity of the Ill*, University of Tennessee Press, Knoxville, pp. 157–166.
223. Pellegrino, E.: 1983, "The Healing Relationship: The Architectonics of Clinical Medicine", in E.E. Shelp (ed.), *The Clinical Encounter*, D. Reidel Publishing Company, Dordrecht, Holland, pp. 153–172.
224. Pellegrino, E.: 1991, "Trust and Distrust in Professional Ethics", in E.D. Pellegrino, R.M. Veatch, and J. Plangan (eds.), *Ethics, Trust, and the Professions: Philosophical and Cultural Aspects*, Georgetown University Press, Washington, D.C., pp. 69–89.
225. Pellegrino, E.D., and Thomasma, D.C.: 1988, *For the Patient's Good*, Oxford University Press, New York.
226. Plato, *The Republic*, Bk. 1, 342
227. Psathas, G.: 1973, *Phenomenological Sociology*, John Wiley & Sons, New York.
228. Psathas, G., and Waksler, F.: 1973, "Essential Features of Face-to-Face Interaction", in G. Psathas (ed.), *Phenomenological Sociology*, John Wiley & Sons, New York, pp. 171–172.
229. Ramsey, P.: 1970, *The Patient as Person*, Yale University Press, New Haven, CT.
230. Rawlinson, M.C.: 1982, "Medicine's Discourse and the Practice of Medicine", in V. Kestenbaum (ed.), *The Humanity of the Ill*, University of Tennessee Press, Knoxville, TN., pp. 69–87.
231. Rawls, J.: 1971, *A Theory of Justice*, Harvard University Press, Cambridge, MA.
232. Reiser, S.J.: 1982, *Medicine and the Reign of Technology*, Cambridge University Press, New York.
233. Reiser, S.J. *et al.*: 1977, *Ethics in Medicine*, MIT Press, Cambridge, MA.

234. Rorty, R.: 1979, *Philosophy and the Mirror of Nature*, Princeton University Press, Princeton.
235. Rosenberg, C.E.: 1987, *The Care of Strangers: The Rise of America's Hospital System*, Basic Books, Inc., New York.
236. Ross, J.W., 1989, "An Ethics of Compassion, A Language of Division: Working out the AIDS Metaphors", in I.B. Corless and M. Pittman-Landgman (eds.), *AIDS: Principles, Practices, and Politics*, Hemisphere Publishing Company, Washington, D.C., pp. 81–95.
237. Rueschemeyer, D.: 1983, "Professional Autonomy and the Social Control of Expertise", in R. Dingwall and P. Lewis (eds.), *The Sociology of the Professions*, The Macmillan Press, London, pp. 38–58.
238. Ruf, H.: 1987, "The Origin of the Debate Over Ontotheology and Deconstruction in the Texts of Wittgenstein and Derrida", in R. Ruf (ed.), *Religion, Ontotheology, and Deconstruction*, Paragon House, New York, pp. 3–42.
239. Sachs, O.: 1985, *The Man Who Mistook His Wife for a Hat and Other Clinical Tales*, Summit Books, New York.
240. Schneidau, H.:, 1976, *Sacred Discontent: The Bible and Western Tradition*, Louisiana State University Press, Baton Rouge.
241. Schumacher, E.F., 1973, *Small is Beautiful*, Harpers, New York.
242. Schutz, A.: 1945, "Some Leading Concepts of Phenomenology", *Social Research* 12, 77–97.
243. Schutz, A.: 1962, *Collected Papers* I, M. Natanson (ed.), Martinus Nijhoff, The Hague.
244. Schutz, A.: 1964, *Collected Papers* II, M. Natanson (ed.), Martinus Nijhoff, The Hague.
245. Schutz, A.: 1966, *Collected Papers* III, M. Natanson (ed.), Martinus Nijhoff, The Hague.
246. Schwartz, M., and Wiggins, O.: 1985, "Science, Humanism, and the Nature of Medical Practice: A Phenomenological View", *Perspectives in Biology and Medicine* 28, 331–334.
247. Scott, R.: 1981, *The Body as Property*, The Viking Press, New York.
248. Sellers, J.: 1989, "Tensions in the Ethics of Interdependence," reprint, pp. 9–10.
249. Sharp, S.: 1988, "The Physician's Obligation to Treat AIDS Patients", *Southern Medical Journal* 81, 1282–1285.
250. Shelp, E.E.: 1984, "Courage: A Neglected Virtue in the Patient-Physician Relationship", *Social Science and Medicine* 18, 351–360.
251. Shelp, E.E.: 1985, *Virtue and Bioethics*, D. Reidel Publishing Company, Dordrecht.
252. Shelp, E.E.: 1986, *Born to Die? Deciding the Fate of Critically Ill Newborns*, The Free Press, New York.
253. Shelp, E.E., and Sunderland, R.H.: 1987, *AIDS and the Church*, Westminster Press, Philadelphia.
254. Shem, S.: 1981, *The House of God*, Dell, New York.
255. Siegler, M.: 1991, "The Secularization of Medical Ethics", *Update* 7 (June), 1–2, 6–8.
256. Silver, G.: 1988, "A Threat to Medicine's Professional Mandate", *The Lancet* 2, 786–787.
257. Smart, N.: 1979, *The Philosophy of Religion*, Oxford University Press, New York.
258. Smith, D.H.: 1985, "Medical Loyalty: Dimensions and Problems of a Rich Idea", in E.E. Shelp (ed.), *Theology and Bioethics*, D. Reidel Publishing Company, Dordrecht, Holland, pp. 267–282.
259. Soe, N.H.: 1962, "Kierkegaard's Doctrine of the Paradox", in H.A. Johnson and N. Thulstrop (eds.), *A Kierkegaard Critique*, Henry Regnery Company, Chicago, pp. 141–155, 207–227.
260. Sontag, S.: 1989, *AIDS and Its Metaphors*, Farrar, Strauss, and Giroux, New York.
261. Spretnak, C.: 1992, *States of Grace: The Recovery of Meaning in the Postmodern Age*, Harper, San Francisco.
262. Starr, P.: 1982, *The Social Transformation of American Medicine*, Basic Books, New York.
263. Stewart, W.L.: 1985, "The Public Perception of Physicians", *Journal of Family Practice* 21, 335–336.
264. Stout, J.: 1988, *Ethics After Babel: The Language of Morals and Their Discontents*, Beacon, New York.

265. Strong, P.M.: 1983, "The Rivals: An Essay on the Sociological Trades", in R. Dingwall and P. Lewis (eds.), *The Sociology of the Professions*, The Macmillan Press, London, pp. 59–77.
266. Suchman, A.L.: 1988, "What Makes the Doctor-Patient Relationship Therapeutic? Exploring the Connexional Dimension of Medical Care", *Annals of Internal Medicine* 108, 125–130.
267. Taylor, M.C.: 1984, *Erring: A Postmodern A/theology*, The University of Chicago Press, Chicago.
268. ten Have, H., Kimsma, G., and Spicker, S.F.: 1990, *The Growth of Medical Knowledge*, Kluwer, Dordrecht.
269. Thomas, L.: 1983, *The Youngest Science: Notes of a Medicine-Watcher*, Viking Press, New York.
270. Tillich, P.: 1948, *The Protestant Era*, The University of Chicago Press, Chicago.
271. Tillich, P.: 1954, *Love, Power, and Justice*, Oxford University Press, New York.
272. Tiryakin, E.: 1984, "The Logic and Limits of Trust", *Society* 22, 89–90.
273. Titmus, R.: 1971, *The Gift Relationship: From Human Blood to Social Policy*, Pantheon, New York.
274. Toombs, S.K.: 1987, "The Meaning of Illness: A Phenomenological Approach to the Patient Physician Relationship", *Journal of Medicine and Philosophy* 12, 219–240.
275. Toombs, S.K.: 1992, *The Meaning of Illness*, Kluwer, Norwell, MA.
276. Toulmin, S.: 1960, *An Examination of the Place of Reason in Ethics*, Cambridge University Press, Cambridge.
277. Toulmin, S.: 1976, "On the Nature of the Physician's Understanding", *Journal of Medicine and Philosophy* 1, 32–50.
278. Toulmin, S.: 1977, "The Meaning of Professionalism: Doctors' Ethics and Biomedical Science", in H.T. Engelhardt, Jr., and D. Callahan (eds.), *Knowledge, Value, and Belief*, Institute of Society, Ethics, and the Life Sciences, Hastings-on-Hudson, NY., pp. 254–278.
279. Tracy, D.: 1986, *Analogical Imagination*, Crossroad, New York.
280. Tracy, D.: 1987, *Plurality and Ambiguity: Hermeneutics, Religion, and Hope*, Harper and Row, San Francisco.
281. Troeltsch, E.: 1977, "What Does 'Essence of Christianity' Mean?", in R. Morgan and M. Pye (eds.), *Ernst Troeltsch: Writings on Theology and Religion*, trans. M. Pye, John Knox Press, Atlanta, pp. 124–181.
282. Veatch, R.M.: 1991, "Is Trust of Professionals a Coherent Concept?", in E.D. Pellegrino, R.M. Veatch, and J. Plangan (eds.), *Ethics, Trust, and the Professions: Philosophical and Cultural Aspects*, Georgetown University Press, Washington, D.C., pp. 159–169.
283. Waldholz, M., and Bodgdanich, W.: 1989, "Doctor-Owned Labs Earn Lavish Profits for a Captive Market", *The Wall Street Journal* (March 1), A1, A6.
284. Walsh, T.G., 1989, "Deconstruction, Countersecularization, and Communicative Action: Prelude to Metaphysics", in H. Ruf (ed.), *Religion, Ontotheology, and Deconstruction*, Paragon House, New York, pp. 114–126.
285. Webster, J.R., Jr.: 1987, "The Physician's Image and Options", *American Journal of Medicine* 83, 123–126.
286. *Webster's New international Dictionary*, 1961, 2nd. Edition, G. and C. Merriam Company, Springfield, MA.
287. Wiebe, P.: 1984, *The Architecture of Religion: A Theoretical Essay*, Trinity University Press, San Antonio.
288. Winquist, C.: 1986, *Epiphanies of Darkness*, Fortress Press, Philadelphia.
289. Winter, G.: 1966, *Elements for a Social Ethic: Scientific and Ethical Perspectives on Social Process*, Macmillan, New York.
290. Wolff, K.: 1973, "Toward Radicalism in Sociology and Every Day", in G. Psathas (ed.), *Phenomenological Sociology*, John Wiley & Sons, New York, pp. 56–57.
291. Wright, P., and Treacher, A.: 1982, *The Problem of Medical Knowledge: Examining the Social Construction of Medicine*, Edinburgh University Press, Edinburgh.

292. Zaner, R.M.: 1961, "Theory of Intersubjectivity: Alfred Schutz", *Social Research* 28, 84–86.
293. Zaner, R.M.: 1973, "Solitude and Sociality: The Critical Foundations of the Social Sciences", in G. Psathas (ed.), *Phenomenological Sociology*, John Wiley & Sons, New York.
294. Zaner, R.M.: 1986, "Embodiment", in J.F. Childress and J. Macquarrie (eds.), *The Westminster Dictionary of Christian Ethics*, The Westminster Press, Philadelphia, pp. 187–190.
295. Zaner, R.M.: 1988, *Ethics and the Clinical Encounter*, Prentice-Hall, New Jersey.
296. Zussman, R.: 1992, *Intensive Care: Medical Ethics and the Medical Profession*, The University of Chicago Press, Chicago.

INDEX

Theology and Medicine

Managing Editor

1. R.M. Green (ed.): *Religion and Sexual Health.* Ethical, Theological and Clinical Perspectives. 1992 ISBN 0-7923-1752-1
2. P.F. Camenisch (ed.): *Religious Methods and Resources in Bioethics.* 1994 ISBN 0-7923-2102-2
3. G.M. McKenney and J.R. Sande (eds.): *Theological Analyses of the Clinical Encounter.* 1994 ISBN 0-7923-2362-9
4. C.S. Campbell and B.A. Lustig (eds.): *Duties to Others.* 1994 ISBN 0-7923-2638-5
5. E.R. DuBois: *The Illusion of Trust.* Toward a Medical Theological Ethics in the Postmodern Age. 1995 ISBN 0-7923-3144-3

KLUWER ACADEMIC PUBLISHERS – DORDRECHT / BOSTON / LONDON